ᴛʜᴇSCIENCE
ᴏғENERGY

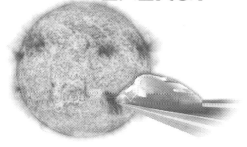

THE SCIENCE
OF ENERGY

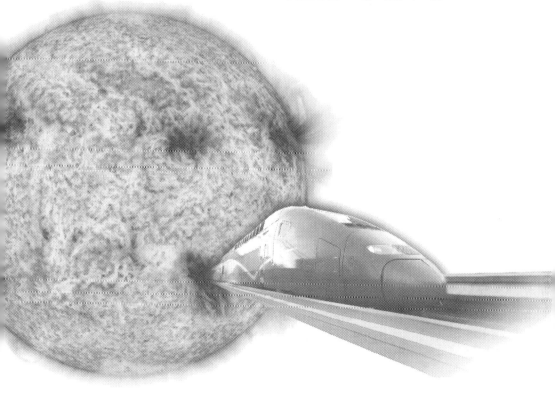

Roger G. Newton
Indiana University, USA

 **World Scientific**

NEW JERSEY · LONDON · SINGAPORE · BEIJING · SHANGHAI · HONG KONG · TAIPEI · CHENNAI

Published by

World Scientific Publishing Co. Pte. Ltd.

5 Toh Tuck Link, Singapore 596224

USA office: 27 Warren Street, Suite 401-402, Hackensack, NJ 07601

UK office: 57 Shelton Street, Covent Garden, London WC2H 9HE

British Library Cataloguing-in-Publication Data
A catalogue record for this book is available from the British Library.

THE SCIENCE OF ENERGY

ISBN-13 978-981-4401-19-7 (pbk)
ISBN-10 981-4401-19-6 (pbk)

Typeset by Stallion Press
Email: enquiries@stallionpress.com

Printed by FuIsland Offset Printing (S) Pte Ltd Singapore

Contents

Introduction   vii

1. Work from Heat and The Basic Laws   1

Energy in Classical Mechanics   6
Heat and the Conservation Law   11
The Second Law of Thermodynamics   13
Einstein's $E = mc^2$ 16
Notes   20

2. Electrical and Chemical Energy   21

Electrical Energy   21
Chemical Energy   27
Notes   30

3. Nuclear Energy   33

Radioactivity   33
Nuclear Fission   36
Nuclear Fusion   39
What Makes the Sun Shine   40
Energy in the stars   43
Notes   45

4. Energy in Quantum Mechanics   47

Discrete Energy Levels   50
Quantum Jumps and Spectra   55
Notes   58

5. Storing and Transporting Energy   61

Long-term Energy Storage   61
Short-term Energy Storage   64
    Flywheels   64
    Pumped Storage   65
    Batteries   65
    Liquid Hydrogen   68
    Fuel Cells   69
Energy Transport   70
Notes   73

6. Energy in the Universe   75

Notes   84

Epilogue   85

Appendix: Research on Energy   87

Battery Technology   87
Other Energy Projects   89

Illustration Credits   93

References and Further Reading   95

Index   97

Introduction

The energy of a system is, therefore, sometimes briefly denoted as the faculty to produce external effects.
Max Planck, Treatise on Thermodynamics, page 41

There are many ways in which modern Western civilization, the so-called developed world, differs from societies on the rest of the planet, such as fully assured property rights, an independent judiciary, a capitalist economic system, efficient transport, etc. But since the industrial revolution two hundred years ago, the overarching achievement of the Western world is surely the substitution of machine-assisted energy for human and animal drudgery. The exchange of machine-powered weapons and transport for individual combat and animal force may not always have altered the outcome in war — the replacement of horses in Napoleon's army by Panzers in Hitler's still led to their defeats in Russia — but the death toll has skyrocketed. Even though Hannibal's victory over the Romans in hand-to-hand combat in 216 BCE may have been the bloodiest one-day battle in history, battles such as those of the Somme and Passchendaele during the First World War, and at Stalingrad and Okinawa in the Second World War, both caused enormous amounts of deaths and injuries, not to mention the atomic bombings of Hiroshima and Nagasaki. But in peacetime mechanization has had lots of positive effects. Machine energy dominates modern agriculture and factory work as well as everyday city living. Plowing, sowing and harvesting machines have taken the place of humans, oxen and horses; washing machines, dishwashers, automatic dryers and vacuum cleaners have replaced both class-based maid-service and dreary housework; and automation has relieved and eased much backbreaking factory labor.

How did we come to understand energy sufficiently to make such a fundamental development possible, and what is the source of this energy? There are, of course, many economic and political issues having to do with access to energy (or the lack of it), such as the short-term power but long-term deleterious social effects that access to enormous oil reserves may have on societies with few other sources of income, and the enormous effects of differential access to sufficient sources of energy on the power relationships between countries. In addition, there are important engineering, social, and political problems such as the dangers of nuclear fission reactors, the disposal of radioactive waste produced by them, and the environmental damage caused by surface coal mining.

But to be clear about the basis of all these questions it is surely good to understand the scientific fundamentals of energy, and that is what this book is about. It will explain both the scientific laws governing the use of energy and the science involved in its utilization, as well as the cosmic history of energy. Written for readers with little or no scientific knowledge or background, it deals with the scientific aspects of energy, such as its definition and underlying laws as well as its various forms, its storage, and its transport. We will be discussing matters of physics and chemistry, as well as some biology.

The first chapter elucidates the origin of the concepts of work and other mechanical energy in Newton's mechanics, the two important laws of thermodynamics, conservation of energy and its degradation for the purpose of actual use, as well as Einstein's enlargement of what constitutes energy by his famous formula $E = mc^2$. The second chapter deals with, and explains, the physics of the two forms of energy that are most frequently utilized: electrical, including radiation, and chemical, in the form of combustion and photosynthesis. If the steam engine made the industrial revolution possible, the engines most of us have the closest contact with nowadays are those powered by gasoline or diesel fuel, the workings of which will be explained. Chapter 3 explains the development of nuclear energy, both in its fission and fusion forms. The latter, as we shall see, is the source of the Sun's radiation that makes life on Earth possible. In Chapter 4 we explain the basic role that quantum theory plays in atomic phenomena concerning energy and the emission of electromagnetic radiation, including light. The fifth chapter deals with the storage of energy, both long-term, as in fossil fuels such as coal, petroleum, and natural gas, and short-term, as in batteries, hydrogen, and fuel cells. This chapter also treats energy transport, by the use of both transmission lines and laser beams, and we discuss the reasons for the high voltages of transmission lines as well as for the ubiquitous employment of alternating current. Finally, the sixth chapter describes the changing forms of energy in the course of the early history of the universe: the 'pure' form in which energy began, and the way all matter grew out of this original form. An appendix lists a sampling of a large number of energy research projects supported by the American government.

When we are done, the reader will understand both the basic scientific ramifications of energy and the fact that it is ultimately the Sun to which we owe almost all the energy we use on the Earth, one of the few exceptions being nuclear energy, the very source that also fuels the Sun.

Acknowledgements

I am happy to acknowledge my indebtedness to my wife Ruth for editorial help and to Kenneth W. Ford as well as an anonymous reader for very useful detailed critiques of the manuscript.

1

Work from Heat and The Basic Laws

The politically and culturally overriding event of the nineteenth century — far outstripping Napoleon's defeats in Russia and at Waterloo, surpassing even the freeing of Russian serfs and the abolition of slavery by Britain and the U.S. — was the industrial revolution in Europe and the United States. The driving machine, both literally and figuratively, of this truly transforming development was the steam engine, made practicable by the Scottish engineer James Watt's invention of a novel condenser-valve.

The use of steam power had first been proposed in a rudimentary fashion in the first century CE by the Greek mathematician and engineer Hero, who lived in Alexandria, a Roman province at the time. Hero also invented a windwheel, the earliest known attempt to harness the energy of the wind. Hero's steam-powered device became well known as the Hero engine, but it never served any practical use. It was not until the seventeenth century that Denis Papin designed a working steam engine. An assistant of the great Dutch physicist Christiaan Huygens, Papin, a Huguenot, left his native France to work in London, later becoming Professor of Mathematics at the University of Marburg. In Papin's engine the steam pressure produced by heated water pushed a piston up a cylinder, which was then cooled, thus replacing the upward pressure with a vacuum when the steam condensed to water and allowed the piston to fall. Rigged to a pump, his device could be used for pumping water, but it was not a practical engine for regular production of mechanical energy, even when gradually improved, until James Watt's pathbreaking innovation.

The valve invented by Watt, a Scottish engineer and inventor working at the University of Glasgow, circumvented the need to constantly heat and cool the entire steam cylinder. His invention enabled this machine to rotate a wheel and to do actual practical work. Of all the engineering inventions of the nineteenth century, this simple idea had by far the most transformative effect on our civilization.[1]

The steam engine gradually not only displaced the cumbersome need for much human and animal labor but was able to perform tasks inconceivable without it. It made the efficient functioning of large factories possible, thereby transforming many European and American cities. It led to the construction of railroads with trains pulled by steam-driven locomotives, greatly shortening land travel, facilitating population growth in the western United States and Canada, and the governance of the subcontinent of India and colonies in Africa.

Huge train stations in big cities, resembling cathedrals in their architectural grandeur, became symbols of the new age.

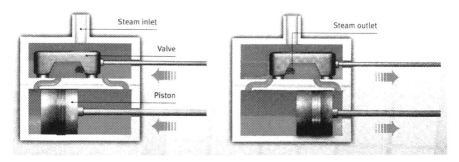

Figure 1.1. The double-acting steam engine with the Watt valve. Pressurized steam from a boiler is piped into the cylinder by means of a sliding valve; exhaust steam leaves the cylinder through the steam outlet.

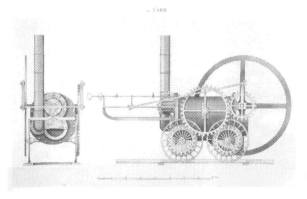

Figure 1.2. The first locomotive, built by Richard Trevithick in 1804, a modern steam locomotive, and a modern high-speed electric locomotive.

Figure 1.3. Victoria Terminus in Mumbai, India

Figure 1.4. Interior of Victoria Terminus in Mumbai, India

Figure 1.5. The interior of Pennsylvania Station in New York before its demolition in 1963.

The steam engine also transformed river travel by means of side-paddle boats as well as long-distance sea travel by enormous steam-propelled ocean liners. Replacing the wind-driven sailing vessels that had been used for centuries, it eventually shortened the duration of a transatlantic voyage from months to less than a week. There was no aspect of Western life that was not fundamentally touched by this engine able to use heat to produce work.

It was only natural that this transformative train of events led scientists to want to understand heat and the conversion of energy from one form to another (although that was not the way matters were seen at the beginning). The new branch of physics that specializes in researching the nature of and discovering the laws governing heat in all its ramifications was called thermodynamics, and during the 1800s it became the most active and innovative field of physics, rivaled only by electromagnetism. It led to a much better understanding of what energy is.

While the word *energy* is of Greek origin, in Aristotle's philosophy the notion of energy existed only in the vaguest form as a force capable of performing work or accomplishing results. There seemed to be no desire or need to make the concept more exact, either in ancient Greece or in medieval Europe. It remained inchoate and vague for another two millennia until the scientific revolution of the sixteenth and seventeenth centuries.

Energy in Classical Mechanics

The form in which energy was first recognized more precisely was as a punch delivered by a moving object. Everyone knows this force: the heavier the object and, especially, the faster it moved, the more powerful the impact. It was called *vis viva*, the living force, defined as the product of the mass of a body multiplied by the square of its velocity: mv^2.

Galileo demonstrated, by rolling a ball down planes inclined at various angles, that the speed of its motion gradually increased with a uniform acceleration which did not depend on its mass. Furthermore, it turns out that the *vis viva* with which the ball arrives at the bottom of a drop is twice its weight times the height of the drop, no matter whether it fell straight down or rolled along an inclined plane. Now since the work[2] required to raise ball to the height H is equal to H times its weight, it follows that the work done to raise the ball is equal to one half the *vis viva* with which it arrives at the bottom. So $mv^2/2$ was subsequently called the *kinetic energy*: the kinetic energy of the ball at the end of its motion is equal to the work performed before the start; the punch exerted by the ball at the bottom, embodied in its kinetic energy, just equaled the work done to begin with. While the ball is resting at the height H, with the work ingested, so to speak, and ready to drop and regurgitate it in the form of kinetic energy, it is said to have *potential energy* equal to that ingested work. During the drop it changes its potential energy to kinetic energy, so that the total amount remains the same. However, the potential energy can also be exchanged for work by attaching the ball with a rope to a weight W resting at the bottom, letting it descend slowly and thereby raising the weight to a height H. The product WH then equals the original work done to raise the ball. Both potential and kinetic energy can always be exchanged for actual work.

Another form of kinetic energy is that of rotation: a heavy rotating wheel has kinetic energy just as a heavy object moving in a straight line. Attaching a string to its rim, with the other end tied to a hanging weight, allows the transformation of this kinetic energy into work, lifting the weight. Conversely, analogous to work being transformed into kinetic energy of straight-line motion by exerting a force on and accelerating an object, so it is converted into rotational kinetic energy by exerting a torque[3] on a wheel. Just as the kinetic energy of an object of mass m moving with velocity v is $mv^2/2$, so the rotational kinetic energy of a wheel rotating at an angular velocity ω — the angle by which it turns per unit time — is $I\omega^2/2$. Here I is its *moment of inertia*, which depends on its mass, its shape, and how the mass is distributed over its shape.[4]

Applications of such conversions have existed for centuries. The ancient Romans as well as the Chinese converted the kinetic energy of falling water — moving liquids and gases of course also contain kinetic energy — into rotational energy by means of paddle wheels, as well as the kinetic energy of wind by windmills (remember Hero of Alexandria) to replace oxen and horses for the heavy work of grinding corn.

Isaac Newton's equations of motion formalized these matters in mathematical language applicable to more general circumstances. During the swing of a pendulum, its bob continually exchanges potential for kinetic energy: at the top of its swing, momentarily at rest, all its energy is in the form of potential energy, while at the bottom it has all been exchanged for kinetic energy. The latter is then gradually expended to buy potential energy as the bob rises. Similarly, when a comet approaches the Sun, it keeps on trading potential

Figure 1.6. A rock-crushing mill illustrated in Georgius Agricola's *De Re Metallica* (1556).

energy for kinetic energy until it turns at its closest approach, when it begins to gain potential energy while losing kinetic energy until it turns again.

However, if you look for 'energy' in Newton's great treatise, *Philosophiae Naturalis Principia Mathematica*, which contains his laws of motion, you will search in vain. The concept of energy is recognized in it only implicitly, and the word 'energy' (or its equivalent, 'energia,' in Latin, the language in which Newton wrote) is never employed. His thinking and argumentation about motion was always more geometrical than algebraic. The great eighteenth-century Swiss-born mathematician, Leonhard Euler, was actually the first to write Newton's second law explicitly as F = ma, the form by which we know it today. He also recognized that Newton's laws of motion implied that any assembly of objects uninfluenced by external forces obey certain specific *conservation laws*. The total energy (the sum of all their kinetic and potential energies), the sum of their momenta, and the sum of their angular momenta are all conserved: these values do not change, no matter how complicated the motion may be. Here the momentum of a particle is defined as the product of its mass m and its velocity, while its angular momentum about a given point is m times its angular velocity about that point. However, we shall concern ourselves only with the energy.

If you have an abstract turn of mind you may well wonder at this point why it is that Newton's laws of motion imply these conservation laws. Would any set of equations of motion, no matter how strange, imply that there are certain quantities which remain constant throughout the motion, even though they depend on the changing dynamical variables? The answer to this question was provided by a German mathematician by the name of Amalie Emmy Noether, whereby hangs a very unusual tale.

The daughter of a mathematician, Emmy Noether was born in 1882 in Erlangen, Bavaria, and decided to study mathematics at the local university. But women were not allowed to enroll as regular students in German universities at the time — to permit women at universities would 'overthrow all academic order,' it was argued — as was also the case in other countries. So special permission by the Academic Senate and by the individual professors of the courses she wanted to take were required for her to pursue her goal. To get her doctorate at the University of Göttingen, and to begin teaching mathematics there at a minimal salary, finally became possible through the strong support of David Hilbert, the greatest mathematician in Europe at the time and a very enlightened man.

When the Nazis came to power in Germany in 1933, Emmy Noether lost her university position because she was Jewish. She left Germany for the United States to become Professor of Mathematics at Bryn Mawr College in Pennsylvania, where she died two years later of a postsurgical infection.

Noether's specialty was algebra, in which she made a considerable name for herself with many important contributions. However, among physicists her reputation rests on her proof of what became well-known as Noether's theorem: whenever the equations of motion are subject to a certain symmetry — or invariance, as it is called — then that implies the existence of a corresponding quantity that remains conserved during the motion. If, for example, the equations remain invariant under translation, i.e., a shift, in space, then momentum is conserved. And specifically, if the equations are invariant under translations and reversal of the time — it matters not if an experiment is performed today or tomorrow, nor does it matter if a film of the motion is run forward or backward — then energy is conserved.

The connection between symmetries and conservation laws that was proved by Emmy Noether became enormously influential in all subsequent fundamental physics. Energy conservation in classical mechanics was not an accidental byproduct of Newton's equations; it was a basic consequence of their structure.

But wait! Is energy really conserved? If you look closely at the motion of a pendulum, you notice that on its second swing it does not rise quite to the same height where it started: the bob has lost some of its potential energy. Indeed, any real pendulum will eventually come to rest, and all its kinetic and potential energies have vanished. Friction prevents mechanical systems from conserving their energy.[5] In the eighteenth century this presented a mystery, which was solved in the nineteenth century when the nature of heat became understood.

Heat and the Conservation Law

What solved the apparent mystery of lost energy in mechanical systems was the recognition that friction produced heat, and heat was also a form of energy. The reason why this fact was not recognized earlier was that heat had for a long time been thought of as a fluid, called caloric, that permeated all matter and had the intrinsic property of flowing from hot to cold regions. However, when friction was demonstrated to be able to melt ice and make iron objects such as gun barrels red hot, the caloric theory was gradually replaced by the kinetic theory of heat: heat was recognized to be nothing but random motion of the molecules making up matter. Heat therefore made up another form of energy. In a certain sense, however, it is an epiphenomenon. From the submicroscopic point of view, there is no heat, only molecular motion. However, from the macroscopic viewpoint, from which we normally see all matter and with which thermodynamics deals, there is a clear distinction between ordered motion of the whole (mechanical energy) and disordered motion of molecular costituents (heat).

The demonstration that in the total amount of energy any physical system, consisting of both mechanical energy (the sum of the kinetic and potential energies of its constituent objects) and its heat, is always conserved was one of the most important achievements of nineteenth century science. This demonstration, however, was accomplished by a variety of means involving great controversies.

One of the men credited with the discovery of the conservation of energy is a German physician by the name of Robert Mayer who served in his youth as surgeon on board a ship during a long voyage to the tropics. He made the crucial observation that the blood of the sailors he treated along the way changed color as the ship entered the tropics. This he interpreted as less work being needed in hot climates to keep the blood in a human body at a constant temperature than in colder ones. By a convoluted chain of reasoning, not all of it correct, he concluded that there must be a fundamental law of nature dictating that the total amount of heat and work is conserved. Convinced that he had discovered an extremely important fact but unable to show his arguments in any scientifically coherent manner, Mayer had great difficulties getting his discovery published. He was a visionary, not a scientist, but he did have the great insight that there is a fundamental entity he called *energy* that came both in the form of heat and in the form of mechanical work, and it is conserved.

Meanwhile, the British physicist James Prescott Joule discovered the same law by other means. An extremely careful and precise experimenter who was convinced that heat was a form of energy, Joule set out to measure exactly how much heat was equivalent to how much work. He arranged for the rotation of a paddle wheel in a container that would warm the water in it by agitating and generating friction, while driving the wheel's rotation by means of a descending weight (see Fig. 1.7). Measuring the rise in the water's temperature and the work required to raise the weight allowed him to determine with great precision how many units of mechanical work were equivalent to a unit of heat.[6] Today, *joule* is the scientific name of a unit of energy.

Another scientist sometimes credited with the dicovery of the law of conservation of energy is the German physician, physicist, and philosopher of science Hermann von Helmholtz. After studying the metabolism of muscle contraction, Helmholtz concluded that during the movement of muscles there was no loss of energy, and that, contrary to many speculative claims, no 'vital forces' were needed. In an influential book entitled *Über die Erhaltung der Kraft (On the Conservation of Force)* he argued that the sciences of mechanics, heat, light, electricity, and magnetism were all related by their reliance on the concept of force, i.e., energy. Whether Helmholtz was aware of

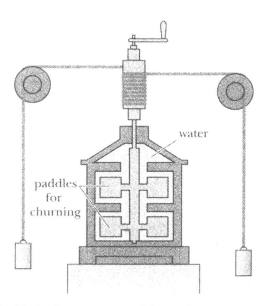

Figure 1.7. The kind of apparatus used by Joule to measure the mechanical equivalent of heat.

Mayer's work was a matter of controversy at the time, but he had, and still retains, a very high reputation in Germany.

'Conservation of energy' is now called *the first law of thermodynamics,* one of the bedrocks of physics. It rules out any possibility of constructing an engine that would run forever, in isolation and without any input of any kind of energy — a *perpetuum mobile,* as such an imaginary gadget would be called. Many inventors over the years nevertheless tried to get patents on such engines, which they thought would solve the world's energy problems.

The Second Law of Thermodynamics

At this point you might think that, even though the first law of thermodynamics does not allow us to get energy for nothing, it also prevents us from losing any of it, and whatever supply of energy the Earth has is forever safe. After all, there is an enormous reservoir of energy in the form of heat in the world's oceans. With sufficiently ingenious technology we should be able to extract that heat, cooling the seas somewhat, and convert it into mechanical energy able to do work. Unfortunately, there is also the *second law of thermodynamics*, just as fundamental as the first.

The man originally responsible for the discovery of this law — indeed considered by some to have been the father of the science of thermodynamics — was a French engineer by the name of Nicolas Léonard Sadi Carnot. Born at the end of the eighteenth century in Paris as a son of a prominent military leader, Sadi Carnot's main motivation was to understand the basic principles underlying the working of all kinds of heat engines that engineers at the time were experimenting with.

An important question was: can the steam engine be improved by substituting another kind of gas or fluid for steam? In order to answer it in general terms applicable to all kinds of heat engines, he envisaged an idealized engine that functioned without friction by simply allowing heat to be transferred from a reservoir[7] at one temperature to another at a lower temperature and in so doing produce mechanical work. The important result of his reasoning, which he presented in a book entitled *Réflexions sur la puissance motrice du feu* (*Reflections on the Motive Power of Fire*) was that the efficiency of any such ideal arrangement — measured as the ratio of the amount of work produced divided by the amount of heat transferred — was a function only of the temperatures of the two reservoirs: the higher the temperature of the hotter reservoir relative to that of the cooler one, the more efficient the engine. Nothing depended on the nature of the fluid employed in the engine, though of course friction reduced its efficiency. His theory made it clear that engines using superheated steam — or other kinds of engines invented much later, such as diesel and gasoline engines — were better than ordinary steam engines because of the higher temperature of their hot reservoirs. Even though Carnot wrote his work before the first law of thermodynamics was clearly established, it opened the way to the discovery of the second law. He died in the cholera epidemic of 1832 in Paris at the age of 36.

The second law is not as easily stated as the first, but one way of saying it is this: there can be no heat engine that functions by simply extracting heat from a reservoir like the sea, sucking heat from it and letting it flow to the engine, if necessary allowing the leftover to leak back into the same reservoir, or into one of the same temperature. Such an engine would be called a '*perpetuum mobile* of the second kind.' Instead, the heat would always have to flow from one reservoir to another of a lower temperature, just as in Carnot's engine. In other words, any heat engine doing work must always produce some waste heat to be ejected into another, cooler reservoir.

As a result, of course, the warmer reservoir gets cooler and the colder one warmer, and the efficiency of the engine gradually decreases until the two reservoirs are at the same temperature and the process comes to a stand-still. If we think of the whole universe as a container full of regions at different temperatures — hot stars and cold spaces — then all the work that can ever be done is obtained by extracting heat from a hot region and transferring it to a colder one. This process tends to equalize the temperatures until eventually all regions are at the same temperature and it becomes impossible to extract work anywhere. This fate of the universe, implied by the second law of thermodynamics, is called its *heat death*. It is not something to worry about in any relevant future because it is many, many billions of years away; it's merely a matter of principle.[8]

From a practical point of view, the second law of thermodynamics implies that even though the total amount of energy in the world is constant and none can be lost, any process of using it by doing work necessarily transforms some of the energy into a useless form — still heat energy, but at such low temperature that no work can be extracted from it. There is no way of avoiding waste heat.

So far we have discussed only two forms of energy, heat and mechanical energy such as work, but there are other forms of equal importance. The fundamental laws of thermodynamics apply to all of them.

Einstein's $E = mc^2$

In the early twentieth century the idea of what counts as energy was once again fundamentally enlarged by the greatest scientist since Isaac Newton.

The year 1905 was a time of magical creativity for Albert Einstein, analogous to Isaac Newton's *anni mirabiles* 1664 to 1666, when he invented calculus and discovered the universal law of gravitation. Born in the city of Ulm in Southern Germany, known for its beautiful Gothic (now) Protestant church called the Münster, Einstein got his early schooling in the same city and chafed under the discipline customary in German schools at the time. He received his higher education in Switzerland at the Eidgenössische Technische Hochschule (ETH) in Zürich. After a year as a school teacher he was happy to find a job in the Swiss Patent Office in Bern examining inventions submitted for their originality and patent-worthiness. As a young clerk in this office, Einstein wrote four fundamental papers (he was just 26 by the time they were published), three of which would trigger revolutions in our understanding of nature as profound as those initiated by Sir Isaac. The first changed our view of the nature of light by introducing the idea of light quanta — later called photons — which began the quantum theory of radiation and will be discussed in the next chapter; the second finally explained the mysterious irregular movement of small particles in water seen under the microscope (called *Brownian motion*) as the result of the movements of the water molecules; the third was the invention of the theory of relativity, which altered our conception of space and time; and the fourth used the results of the third to introduce $E = mc^2$ and the notion of mass as a form of energy.

Though not much noticed and even less understood by other physicists, his four papers did lead to a junior professorship at the

University of Zürich, followed by a full Professorship at the University of Prague in 1911. There was, however, one well-known physicist who did pay attention to and appreciate the originality and importance of Einstein's work: Max Planck, the man who had taken the first step that eventually led to the quantum theory (see Chapter 4), and he arranged for him to be appointed as Director of the Kaiser Wilhelm Institute for Physics in Berlin, Germany. This was where Einstein was when he succeeded in 1915, after a great deal of struggling with unfamiliar mathematics, in constructing his theory of gravity called the general theory of relativity (see Chapter 6), a fundamental break with Isaac Newton. When an expedition to West Africa, led in 1919 by the British astronomer Arthur Stanley Eddington to observe a solar eclipse, was able to measure the precise amount by which the Sun's gravity influenced stellar light (which could be measured only during an eclipse), their result verified one of the predictions of the general theory of relativity that differed from what Newton's law implied and Einstein instantly became world famous as the scientist who had proved Newton wrong. Travelling and lecturing everywhere, he was treated by the press like a film star.

Celebrated as he was elsewhere, in his native Germany Einstein was vilified as a Jew and attacked by the growing Nazi movement, even among some of his colleagues. Though not a conventionally religious man, he was never ashamed of being Jewish or tried to hide it. He also identified himself with certain strong beliefs, such as pacifism — he was a great admirer of Gandhi and his methods — as well as, to some extent, Zionism (though he never contemplated actually moving to Palestine). His aversion to nationalism notwithstanding, he publicly defended the establishment of the state of Israel. In fact, he so strongly defended it that after its first president, Chaim Weizmann, died in 1952, Einstein was officially offered the position as his successor, which he declined.

All his life Einstein was strongly averse to conventionality and independent of opinions around him. During the First World War he was one of the few prominent German professors who refused to sign the notorious "Manifesto of the 93" that denied any German guilt in starting the war and misbehavior of German troops in Belgium. In his later years he often dressed unconventionally and most photographs showed him with tousled hair. The same independence of thought that drove his physics to revolutionary innovations characterized his daily behavior. He loved music and played the violin passably well, Mozart being his favorite composer. At an early age he married a fellow student and had three children with her: a daughter named Lieserl and two sons, Eduard and Hans Albert. The marriage, however, turned out quite unhappy and ended in divorce, after which he married a cousin and they remained together until her death in 1936.

In 1933, when Hitler came to power and Nazi thugs broke into his house near Berlin while he was visiting the United States, he was advised by his friends that it would not be safe for him to return. Einstein then decided to accept a professorship at the Institute for Advanced Study in Princeton, New Jersey, which he had been offered even before its actual founding by Abraham Flexner. This is where he remained until his death.

As a consequence of his newly established theory of relativity, in the fourth of his pathbreaking papers mentioned above, Albert Einstein came to the conclusion that when an object radiates an amount of energy E, its mass is automatically diminished by the amount $m = E/c^2$, where c is the velocity of light. Therefore, he concluded, the mass of any object is a measure of part of its energy; changing the former by an amount m results in a change of the latter by an amount E, and the two are related via the equation[9] $E = mc^2$. Just as Joule's experiment had both extended the meaning of energy to include heat and established its conservation while measuring the ratio of units of heat to mechanical energy, so Einstein extended the meaning of energy once more to include mass, while $E = mc^2$ specified the ratio of mass units to energy units needed to state the conservation law precisely. If Joule measured the mechanical equivalent of heat, Einstein postulated the energy equivalent of mass.

Einstein's equation implied at the same time that the energy of a freely moving particle of mass m had to include not only its kinetic energy owing to its motion but also a 'rest energy' of the amount $E = mc^2$. In other words, even when not moving at all, a particle with no external forces acting on it — and hence no potential energy — still had that large[10] amount of energy expressed by the formula $E = mc^2$. In practical applications this additional energy played no significant role because no known energy transformation from one kind to another changed the masses of any of the particles involved, so the same additional large number appeared on both sides of any equation expressing energy conservation. This, however, would change within the next fifty years with new developments and discoveries in nuclear physics, as we shall see. Einstein did not anticipate that mass could indeed be converted into kinetic energy by means of nuclear reactions, nor that kinetic energy as well as radiation could be converted into mass by the creation of particles, as was later discovered by experiments using large accelerators. But his new law would turn out to be fundamental to much of subsequent physics.

In 1939, after nuclear fission was discovered in Germany (see Chapter 3) and recognized by knowledgeable physicists as possibly enabling the construction of a devastating explosive device in Hitler's hands, Einstein, the passionate pacifist, was persuaded by a small group of colleagues and friends to send a letter to President Roosevelt, alerting him to the danger and urging him to authorize a strong program of research into the possibility of a nuclear-fission weapon. The eventual result of this letter was the Manhattan Project: the construction of the vast laboratory at Los Alamos, New Mexico, and what became known as the atomic bomb. Its use — deplored by Einstein — would end the Second World War.

The greatest scientist since Isaac Newton died in Princeton in 1955.[11]

Notes

[1] Watt's memory is honored today by the use of his name as the unit of energy flow.

[2] Work is defined as the distance over which an object is moved multiplied by the force that needs to be overcome.

[3] Torque is defined as the product of the distance from the center at which a force is applied multiplied by the strength of that force at a right angle to the radius.

[4] The moment of inertia of a solid cylindrical wheel of mass m and radius r is given by $mr^2/2$.

[5] Newton's equations of motion were an idealized abstraction designed to be valid in frictionless free space; friction had to be added as a minor modification.

[6] The amount of work, according to its definition, is measured in terms of force-units times distance-units, while the amount of heat is measured in terms of the rise in temperature of one gram of water.

[7] The word *reservoir* simply means a large source of heat at a given temperature.

[8] It is also now recognized that the presence of gravitation changes the situation in a fundamental way.

[9] For a detailed history of Einstein's discovery of his famous equation see the article by Eugene Hecht, "How Einstein confirmed $E_0 = mc^2$ ", *American Journal of Physics*, 79, June 2011, page 591.

[10] The amount of energy is large, even if the mass is very small, because the velocity of light is an enormous number, and even more so when squared.

[11] Among the numerous Einstein biographies I especially recommend the one by Walter Isaacson.

2

Electrical and Chemical Energy

Electrical Energy

We are all, of course, familiar with energy in the form of electricity flowing through conducting wires made of copper or aluminum (or of gold, when excellent conduction is crucial). When we plug a lamp cord in a wall outlet and flick a switch, the light goes on. The incandescent light bulb invented by Thomas Edison near the end of the nineteenth century emblazons the nights of all the cities of the industrial world. But the energy flowing through the wire into the bulb is not all converted into light; in fact, most of it is turned into heat, as you can feel when you touch the hot bulb. Whenever an electric current flows through a wire at room temperature, some heat is produced — exceptions called *superconductivity* occur only at extremely low temperatures. If the wire is thin enough, the metal (in a light bulb it's usually tungsten) gets so hot that it glows. The light produced is really a byproduct of the heat production. Some kinds of incandescent light bulbs, called halogen bulbs, are filled with an inert gas and a small amount of iodine or bromine (these are called halogens). Together with the tungsten filament, this produces a chemical reaction that increases the lifetime of the filament and allows it to be heated to a much higher temperature. The result is that the color of the emitted light is shifted more toward the blue end of the spectrum, which increases the bulb's efficiency. Because of the heat produced, all incandescent light bulbs are extremely wasteful. That's why other kinds of bulbs, such as fluorescent and light-emitting diode

(LED) bulbs, are more efficient: they produce light without a glowing wire and thus create much less waste heat.

In fluorescent bulbs filled with mercury vapor, electrons that are emitted by an electrode collide with the mercury atoms. The electric current thus excites them, and in their subsequent de-excitation they emit (see Chapter 4) ultraviolet radiation that produces fluorescence in the coating on the inner surface of the bulb. Fluorescence is the visible light emitted, without accompanying heat. Light-emitting diodes (LEDs) make use of a fundamental property of semiconductors[12] that emit light when an electric field excites and rearranges their constituent electrons. This phenomenon is called electroluminescence, and it produces no heat. However, LEDs can be made only in the form of very weak individual bulbs, so to create a strong light requires the assembly of many of them. But a little night light made that way, when run day and night continuously for about six weeks, uses only as much electricity as a 30 watt incandescent bulb does in one hour.

The flow of energy — physicists call the rate of energy flow *power* — is measured in terms of watts. These are units of energy per unit time, which in electrical terms is proportional to the product of the voltage — 110 volts for American household wiring but 220 volts in many other countries — multiplied by the amount of current flowing through the bulb, measured in terms of ampères. The total amount of electrical energy used, for which you get a monthly bill, is the rate of energy flow multiplied by the time it is turned on, watt seconds, or more practically, kilowatt hours, abbreviated as kWh.

From a practical point of view, the most important form of energy is always the mechanical of either work or kinetic energy: heavy lifting or else accelerating a movable object such as a car, a train, or an airplane. One way of transforming electrical energy into such mechanical energy is by first producing heat, which can then be transformed into work in various ways such as by means of a steam engine. There is, however, also a direct way, by means of the electrical engine.[13]

An electrical engine, also loosely called an electric motor, uses the magnetism produced by an alternating current to produce a torque on a magnet turning on an axle. Discovered first by the American physicist Joseph Henry and independently a year later by the English scientist Michael Faraday, who published the discovery first, this effect is called electromagnetic induction and is governed by what was named 'Faraday's law.' It directly transforms electrical energy into mechanical energy in the form of rotation. Most of the work performed in today's factories is done by means of electric motors rather than steam engines. Such motors are much more flexible and come in a vast variety of sizes, including tiny ones used to drive household appliances and even watches.

Conversely, a varying magnetic field produces a current in a wire, an effect discovered by the French physicist André Marie Ampère, after whom the strength of an electric current was named. Thus a rotating magnet, or else a rotating wire loop in a fixed magnetic field, can be employed to convert mechanical energy into electric energy: this is called a dynamo or an electric generator. Such machines are used on a very large scale, for example, for converting the kinetic energy contained in falling water into electrical energy by diverting part of a descending torrent into turbines that drive generators. Hydroelectric power stations are located at natural waterfalls such as Niagara Falls, on the border between New York state and Canada, as well as at large water dams like the Three Gorges Dam in China and the Hoover Dam in Colorado. The station at the Three Gorges Dam produces as much as 18,300 megawatts[14] of electric power.

Most of our everyday electricity, however, is produced by dynamos turned by heat-driven engines, which in turn are fueled by coal, diesel oil, gasoline, natural gas, or, indirectly, by nuclear fission. All such conversions of one form of energy into another inevitably involve losses and are inherently inefficient. There is no avoiding the second law of thermodynamics.[15]

Electromagnetic energy is not confined to current-carrying wires. According to Maxwell's equations, which govern all electromagnetic phenomena, including light and other electromagnetic waves such as radio waves, microwaves, infrared radiation, and ultra-short wave X-rays and gamma radiation, all electromagnetic fields in space carry energy. It was another of Michael Faraday's ingenious contributions to have imagined that the electromagnetic force — the forces exerted by magnets on one another and on currents as well as those exerted by electric charges and currents on each other and on magnets — is transmitted by means of a condition of free space which came to be called an *electromagnetic field*. All the properties of this field were later encapsulated in mathematical form by the Scotsman James Clerk Maxwell, and it turned out (as was discovered by the German physicist Heinrich Hertz) that what his equations described included what we now call electromagnetic radiation. The various colors into which Newton had discovered sunlight could be decomposed into by means of a prism (see Fig. 2.2) are nothing but oscillations at different frequencies of the electromagnetic field.

What is more, this field carried energy! You might say that the fact that the electromagnetic field carries energy is convincing evidence of its real existence; it might otherwise be considered a purely theoretical construct. Light falling on a mirror exerts a tiny amount of pressure that is, in principle, able to accelerate it and thus be converted into mechanical energy. Miniscule as this effect may be, the total rate at which energy flows continuously in the form of light as well as infrared and ultraviolet radiation from the Sun to the Earth is enormous: approximately 1.75×10^{17} (175 followed by fifteen zeros) watts, about 30% of which is reflected back into space. It accounts for almost all the energy we and all organic life around us use for much of our existence. The source of that energy to heat the Sun and generate all this life-giving radiation, as well as why it gets to us in the form of the colors that Newton saw by means of his prism, will be discussed in the next two chapters.

Figure 2.1. Electric generators made in Hungary for the hydroelectric station in Iolotan on the Murghab River in 1909.

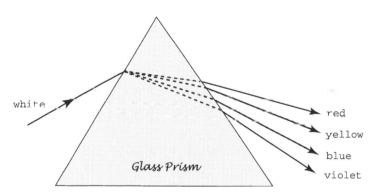

Figure 2.2. The decomposition of white light by refraction in a prism discovered by Isaac Newton.

The electromagnetic energy of light can be converted into the form flowing in wires by means of the photoelectric effect. This phenomenon, discovered in the late nineteenth century by the German physicist Philipp Lenard (who later became an ardent Nazi), consists of a release of electrons when light strikes the surface of a metal. The strange, counterintuitive relations between the amount of light and its color on the one hand and the number and energy of the electrons emitted on the other was explained for the first time by Einstein. Of the two premises on which he based his explanation[16] one was traditional and the other revolutionary: conservation of energy and the idea that the energy of light is 'quantized,' i.e., it always comes in the form of 'quanta' (later called photons), each carrying an amount of energy equal to the frequency of the light multiplied by a constant that had been introduced by Max Planck. Each photon striking the metal surface, he assumed, liberated one electron and made it available for conduction, which explained why the number of electrons was proportional to the intensity of the light shining on it. By energy conservation, the energy of each freed electron increased with that of the liberating photon, and hence with the frequency of the incident light, the observed phenomenon that seemed so mysterious.

Today, this conversion of the radiated form of electromagnetic energy into the motion of electrons takes place commercially in the familiar solar panels, which convert sunlight directly into wire-conducted electricity without any wasteful intermediary. Employing either the photoelectric effect or the photovoltaic effect (a phenomenon of semiconductors, in a sense the converse of LEDs), they can be deployed on a small scale individually, or on a big scale in solar power plants consisting of large arrays of solar panels. Other solar power plants use big parabolic mirrors to concentrate sunlight on a fluid at their focal point heating it up and converting the heat into mechanical energy as in a conventional power plant. Suitable for areas with long periods of bright sunshine, such stations avoid the inefficiency of producing electricity via heat engines, but their overall efficiency is not very large. The price of solar panels, however, is beginning to drop steadily.[17] Individual solar panels can also be used for interior lighting and other purposes in individual houses not accessible to the electricity grid.

Chemical Energy

The most important form of energy in our everyday lives is chemical energy, the heat released in certain chemical reactions, especially those involving oxygen that are called combustion. The volatility of these reactions ranges from the slow form of burning to the very rapid explosive form.

The history of understanding the chemistry of combustion was tortuous, as for a long time it was believed to require no air but to produce a gas called phlogiston. It was the English Presbyterian theologian and chemist Joseph Priestley who reluctantly recognized that, on the contrary, the presence of a gas that he called 'dephlogistinated air' — subsequently named oxygen by the great French chemist Antoine Laurent Lavoisier — was needed for all combustion as well as respiration. A vocal supporter of the French Revolution, Priestley became politically extremely unpopular in his home land and emigrated to the United States, settling in Northumberland, Pennsylvania; Lavoisier fell victim to the Revolution, and his life was cut short under the guillotine.

The fundamental physical reason why these chemical reactions produce heat is the fact that the bonds that hold the atoms in molecules of matter together vary greatly in strength. For example, consider a molecule of water, H_2O.[18] The force binding the atoms of the molecule of water to one another being particularly strong, it requires a large amount of (mechanical) energy to rip the molecule apart into its constituent atoms. This implies, conversely, that when oxygen combines with hydrogen to form water, a lot of energy is available for release in the form of molecular motion, i.e., heat.[19]

When a material such as wood, coal, oil, or natural gas that is rich in hydrocarbons, i.e., chemical compounds made of carbon and hydrogen, is combined with air (21% of which is oxygen) at a sufficiently high temperature — so that their rapid motion brings the molecules closely together as they collide — combustion occurs.

Oxygen atoms from the air combine with hydrogen atoms from the compound — these are not nearly as tightly bound together as the atoms in water — to form water molecules, releasing energy in the form of heat as well as producing carbon dioxide, CO_2. Different materials require various ignition temperatures, release very different amounts of heat, and at different rates: wood burns slowly, natural gas much faster, producing much more heat per unit weight. Note that the release of CO_2 is an inevitable byproduct of the combustion of chemical fuels containing carbon.

So the conversion of chemical energy into work usually proceeds via the production of heat. This heat can then be transformed into mechanical energy by such means as the steam engine, or more directly by means of internal combustion engines, which are specifically designed to control the explosive power of combustible hydrocarbons. The first relevant design of such an engine, the gas turbine (patented in 1791), was invented by the English coal master John Barber. While he had intended it to be used for a 'horseless carriage,' it was never practical for that purpose. Instead, it became very important about a century later for driving large conveyances such as ocean-going ships and airplanes. The gasoline engine that turned out to drive the horseless carriage was invented by the Belgian engineer Etienne Lenoir in 1859 and improved by Gottlieb Daimler and Wilhelm Maybach (see Fig. 2.3). Later, Rudolf Diesel invented what became known as the diesel engine, which differs from the gasoline engine by not requiring an internal spark for its ignition. Whereas in the gasoline engine a mixture of air and fuel enters the cylinder and is then ignited by a spark, in the diesel engine air enters and is heated by compression, so when fuel is injected it ignites by itself, which is why diesel engines need no spark plugs.

These new engines enabled the development of the automobile near the end of the nineteenth century. We are all aware, of course, of the enormous effect of the ubiquitous use of cars and trucks, both on our daily convenience and on the environment. Apart from the

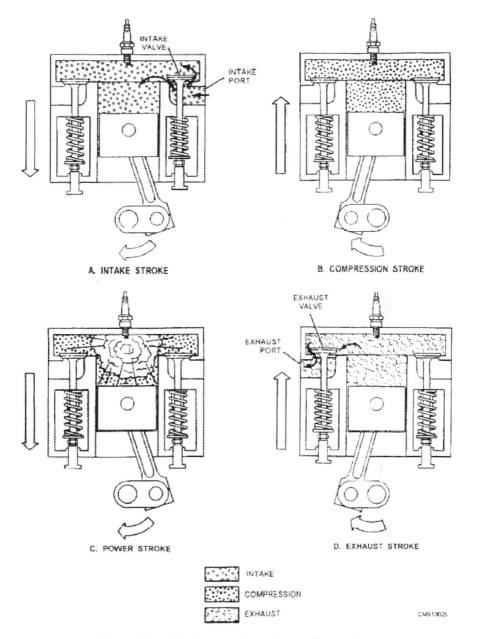

Figure 2.3. The four-stroke-cycle gasoline engine.

unavoidable emission of carbon dioxide in any chemical combustion of carbon-containing materials, the huge number of automobiles all over the world has led to an unquenchable thirst for hydrocarbons in the form of gasoline and diesel fuels.

Conversely, while there is no known process that directly converts mechanical into chemical energy, there is an extremely common process that transforms electrical energy in the form of light into an enormous store of chemical energy: *photosynthesis*. Its chemistry is expressed in the formula: carbon dioxide + water + light → carbohydrate + oxygen.[20] This is the process that drives and sustains almost all of our plant life and at the same time releases oxygen into the atmosphere while reducing its content of carbon dioxide. It is, literally, the life-producing and life-sustaining role of the sunlight that falls on the Earth. All green plants and blue-green algae contain chlorophyll, the chemical that is responsible for photosynthesis. Since all animal-life (including all bacteria, except the so-called anaerobic ones) requires oxygen for its existence, animal and plant-life produces a mutually sustaining cycle of oxygen production in plants and its inhalation by animals, enabled by the energy flow from the Sun. The same cycle simultaneously has green plants absorb the carbon dioxide emitted by breathing animals.

We have seen in this chapter that the chemical form of energy, one of the two forms we use constantly in our daily life as well as the most common form in which we utilize the vast influx of energy from the sun, is transformed into actual work by being first turned into heat, which is then used to perform work. It's not a very efficient way of proceeding, but this is the state of our present technological development.

Notes

12 Semiconductors are materials whose electrical conductivity lies between actual conductors and insulators.

[13] A word about customary terminology: the term *engine* means a machine that converts other forms of energy into mechanical energy.

[14] Mega means million.

[15] For old-style coal-fired power plants, the amount of electrical energy produced is less than one third of the source energy consumed.

[16] This is the paper for which Einstein won his Nobel Prize. It not only solved the riddle of the photoelectric effect, but it initiated the quantum theory of radiation.

[17] For much more information about solar panels see the book by W. Palz, *Power for the World.*

[18] The capitalized letters of the symbol for a chemical compound such as H_2O indicate the elements it consists of, and the subscripts show the number of elementary atoms in each of its molecules, as you may remember from your high-school chemistry class.

[19] Chemists call such reactions *exoergic* or *exothermic.*

[20] Carbohydrates are compounds made up of carbon, hydrogen, and oxygen.

3

Nuclear Energy

The existence of nuclear energy was not discovered until the middle of the twentieth century, when the atomic nucleus became the subject of intense interest for physicists. However, implicitly, its recognition really started near the end of the nineteenth century with the discovery of radioactivity.

Radioactivity

It all began with Ernest Rutherford, who is regarded by many as the greatest experimenter of the twentieth century. Born and educated in New Zealand, Rutherford left for post-graduate education at Cambridge University in England, never to return except to marry the woman he loved and to bring her to England. After a first appointment at McGill University in Canada, he moved to Manchester, England to set up a laboratory that quickly became world famous, after which he moved to Cambridge to take over the Cavendish Laboratory, leading it for the rest of his life. In 1908 he won the Nobel Prize in Chemistry — a strange irony for a man who regarded physics as the only real science, pronouncing all others no better than stamp collecting — and died in 1937 as Baron Rutherford of Nelson (the name of the town close to which he was born), buried near Sir Isaac Newton at Westminster Abbey.

Rutherford's first contribution came after the French physicist Henri Becquerel had accidentally discovered in 1896 that radium emitted a penetrating kind of radiation that darkened photographic plates. This radiation emitted by radium and other heavy radioactive elements that had meanwhile been discovered had been seen to contain three different components, and Rutherford found that these consisted of doubly positively charged helium ions, which he called alpha rays; electrons, which he called beta rays; and very short-wave electromagnetic radiation, which he called gamma rays.

For a while many physicists thought that these radioactive emissions violated the conservation of energy. How could atoms simply continue to emit energetic radiation? The puzzle was resolved when Rutherford found that the radioactivity of any given piece of material slowly decreased, each radioactive element having a characteristic 'half-life' after which the intensity of emitted radiation dropped by 50%, meaning that only half of the radioactive atoms in a given sample were left. The rest turned into other atoms — almost like the old alchemy — that were either stable or had their own, distinct kind of radioactivity. The lengths of these half-lives vary from fractions of seconds to thousands of years, and the radioactivity of the long-lived elements seemed not to decrease at all, appearing to generate energy forever.

Rutherford's greatest discovery was that atoms are almost completely empty, except for a tiny, positively charged speck at the center — like a fly in a cathedral, Rutherford said in astonishment — that contained essentially all of its mass: the nucleus. The atom is kept electrically neutral by a number of very light electrons circulating around this nucleus.

Years later, there was another occasion on which radioactivity appeared to violate the conservation of energy. For alpha radioactivity, things were relatively simple: before a given nucleus of mass m decayed, it had a known energy $E = mc^2$; after its decay, there were two particles with known mass: the daughter nucleus and the emitted alpha particle (a helium nucleus). The conservation laws of energy and momentum then uniquely determine the energy of the emitted alpha particle. The same should be true in the case of beta radioactivity, where the emitted particle is an electron instead of an alpha particle. However, detailed experiments showed that, in contrast to the case of alpha-radioactivity, everything else being the same, the emitted electrons did not always have the same energy! Even Niels Bohr, the father of the quantum theory of atoms, was ready to ditch the sacred conservation law of energy. However, Wolfgang Pauli, the young genius who had already made his mark by discovering how to deal with the strange 'spin' of the electron, suggested that there was another explanation: a beta-decaying nucleus emitted not just an electron, but also a hitherto unknown, electrically neutral, massless (or almost massless) particle, which remained undetected. The name given to this new particle, invented by the Italian Enrico Fermi, was *neutrino*, the 'little neutral one.' The energy of the emitted electrons would then vary from case to case, depending on that of the fugitive neutrino. Electrically neutral and interacting only extremely weakly with other particles, the neutrino took many years to be directly detected, but it was eventually found. Pauli was right, and energy conservation was saved.

Radioactive elements in the interior of the Earth obviously create heat, since they emit particles with kinetic energy, which collide with the atoms of surrounding material, tending to warm the Earth. This source of heat in the lower depths of the ground, called geothermal energy, is brought to the surface by volcanos and hot springs. The latter have been exploited for warming baths and even floor heating since ancient times. Today it is used for generating electricity, for example in power plants at The Geysers, a large geothermal field in California.

The newly discovered source of heat in the Earth had a quite unexpected effect on an unrelated, at that time controversial branch of science, Darwin's theory of evolution. Before the discovery of radioactivity, the highly esteemed physicist Lord Kelvin had estimated the age of the Earth by calculating its rate of cooling from the time of its creation in a hot, molten state as about 100 million years, and certainly no more than 400 million years, which was much too short a time for Darwin's evolution to spawn *Homo Sapiens*. This was a scientific objection to be taken seriously. The discovery of radioactivity, however, drastically changed the estimate of how long the Earth has been in existence, since it slowed down its rate of cooling. Its age is now agreed to be about 4.5 billion years, more than enough time for natural selection to accomplish its Darwinian task.

Radioactivity was the first form in which the importance of nuclear energy manifested itself. Two other forms turned out, from a practical point of view, to be much more powerful: nuclear fission and nuclear fusion.

Nuclear Fission

One source of the hitherto unknown kind of energy appearing on the scene — nuclear energy — is the fact that the nuclei of certain heavy atoms are unstable: they tend to break up into pieces consisting of lighter nuclei. Two German chemists, Otto Hahn and Fritz Strassmann, discovered this accidentally in 1938 when they were trying to produce very heavy 'transuranic elements' by bombarding uranium with slow neutrons. Enrico Fermi, the Italian physicist who invented the name 'neutrino,' had found in 1934 that slow neutrons were much more effective than fast ones in artificially producing radioactive isotopes of certain elements. The neutrons that Hahn and Strassmann used for their experiment were obtained by shooting alpha particles at beryllium, generating ^{13}C plus neutrons.[21]

Mysteriously, and to their great surprise, the two experimenters obtained the much lighter element barium. When they mailed their strange results to their long-time collaborator, the physicist Lise Meitner, who had recently been forced to flee Nazi Germany as a Jew, she and her nephew Otto Frisch quickly found the explanation: the nucleus of the isotope of uranium they had formed by means of their neutron bombardment had broken into pieces, an event never seen before and not even imagined by the chemists Hahn and Strassmann. Eventually more than twenty transuranic elements were later found to exist, the first and lightest ones being neptunium, number 93 in the periodic table, and plutonium, number 94 (uranium is number 92).

Specifically, when a slowly moving neutron collides with the nucleus of an atom of the rare isotope of uranium, ^{235}U (this isotope makes up only 0.7% of naturally mined uranium, which consists mostly of ^{238}U), the uranium nucleus absorbs it. However, the newly formed nucleus ^{236}U is unstable and breaks up — it *fissions* — into two fragments with atomic weights of about 133 and 100, respectively (the exact fissioning breakup is not predictable), plus two or three slow neutrons and a certain amount of electromagnetic radiation. Another fissionable element is ^{239}Pu, a radioactive isotope of plutonium, which cannot be found in nature but can be produced by bombarding ^{238}U with neutrons.[22]

The extra amount of energy released in the form of radiation plus kinetic energy of the motion of the two heavy fission fragments of ^{235}U — and analogously for ^{239}Pu — results from the fact that the total mass of the fission fragments plus the released neutrons is slightly smaller than the initial mass of the ^{235}U nucleus plus the incident neutron, and this mass difference is converted into radiation and kinetic energy according to Einstein's $E = mc^2$. The crucial additional fact is that since each such fission event, caused by the collision of a single neutron with a ^{235}U nucleus, releases more than one slow neutron which are then available to collide with other ^{235}U nuclei to cause fission, the result is a chain reaction — like a snow-ball effect. If more than enough ^{235}U (the lower limit is called a 'critical mass') is sufficiently close together — so that the emitted slow neutrons do not either escape or else first produce enough heat by collisions to expand the material, putting many other ^{235}U nuclei out of reach — to sustain the avalanche, there may be an enormous explosion. Even though the energy released in each individual fission event is quite small, the number of atoms involved is so huge that such an explosion can be vastly larger than any practical chemical explosion. It was this fact that sent a shiver of apprehension through all knowledgeable physicists around the world when they learned about the fission discovery in Hitler's Germany.[23]

Alternatively, if most of the neutrons emitted in each fission event are sponged up in some suitable absorbing material, the chain reaction can be tamed and the resulting machine, first called a pile but now named a *nuclear reactor*, produces heat available for work in the usual way. Such a reactor was successfully constructed for the first time in 1942 at the University of Chicago under the leadership of Enrico Fermi (who had fled fascist Italy and emigrated to the United States), using graphite to slow down the emitted neutrons and cadmium-coated rods as the neutron-absorbing material to prevent an explosion. Werner Heisenberg, who was in charge of Germany's nuclear program during the war, had attempted the same task but never succeeded.

If the chain reaction of ^{235}U in a nuclear reactor is tamed by surrounding or mixing the ^{235}U core with ^{238}U, the emerging neutrons will convert some of the ^{238}U into ^{239}Pu. With enough ^{238}U admixed such a reactor produces more fissile material than it consumes and it is called a *breeder reactor*.

In contrast to most chemical combustions, nuclear energy produces no CO_2. On the other hand, some of the fission fragments are radioactive, with long half-lives. In other words, nuclear reactors usually produce a large amount of waste that stays radioactive for many years.

Nuclear Fusion

There is another kind of nuclear reaction available for the release of energy, called *fusion*. When a proton (a hydrogen nucleus) gets close enough to a neutron for the strongly attractive, short-range nuclear force to take effect, the two particles bind together to form the nucleus of an isotope of hydrogen, ^{2}H, called deuterium.[24] The mass of this nucleus, called a deuteron, is slightly less than the sum of the masses of the initial proton and neutron, so that there is some energy available by Einstein's formula for conversion into kinetic energy.[25] When a neutron approaches a deuteron closely enough, the same thing happens and they form the nucleus (called a triton) of another hydrogen isotope, ^{3}H, named tritium; and when two deuterons energetically collide, they may form a triton plus a leftover proton or else the nucleus of ^{3}He, an isotope of helium, plus a neutron.

The kinetic energy released in each such fusion reaction (and in many others among light nuclei rather than the heavy nuclei involved in fission reactions) is smaller than in individual fission events, but the total energy released is in principle unlimited, depending only on the total amount of material available for fusion.

In contrast to fission reactions, fusion does not form a snowballing chain reaction, and it requires that the initial particles have large kinetic energy for them to approach one another sufficiently closely. This would naturally be the case if the particles found themselves in an environment of extremely high temperature, or it can be artificially achieved by an explosion. Such a *thermonuclear reaction* of the first kind is the energy-mechanism that keeps the Sun and the stars shining; the second kind is the technique employed in a hydrogen bomb, where the trigger used is a fission explosion.

To generate mechanical or electrical energy by controlled fusion reactions for practical purposes would be extremely useful. The initial ingredients are abundantly available (while it takes energy to obtain the needed heavy hydrogen from water, the gain from fusion is very much larger), and neither CO_2 nor large amounts of radioactive debris would be created. Fusion reactors, in contrast to those based on fission, are also not subject to the danger of accidental explosions due to runaway chain reactions. However, the technique of controlling and containing the needed extremely high temperature and resulting pressure has not yet been perfected in ongoing experiments, and fusion reactors are things of the future.

Note that the principle underlying kinetic energy production by nuclear fission as well as fusion is the same as that of chemical combustion: the end products are particles — lighter nuclei in the case of fission, heavier nuclei such as alpha particles in fusion, and water molecules in the case of combustion — that are particularly tightly bound together, thus freeing some energy to be converted into the kinetic form. The fact that in the nuclear instance this manifests itself in a measurable mass difference indicates the much larger kinetic energy production than in the instance of chemical combustion, where the mass difference between a molecule of water and its separate constituent atoms of oxygen and hydrogen is much too small to be measurable.

As was already mentioned, of all the sources of energy we use on the Earth, nuclear is also the fuel that heats the Sun. Let's look at this process, so vital to us on Earth, in a little more detail.

What Makes the Sun Shine

The Sun has existed for billions of years, and the small fraction of its heat and light radiation that has shone on the Earth has sustained life on this planet since its beginning. However, it took the development of nuclear physics to understand where the enormous amount of energy that our nearest star continually sends to us free of charge originates.

Because the temperature of the interior of the Sun is about[26] (16×10^6) K, it is hot enough for thermonuclear reactions to take place, and the German physicists Hans Bethe and Carl Friedrich von Weizsäcker proposed what is known as the CNO cycle of reactions as the source of the Sun's heat. Here is the sequence of fusion reactions that make up this cycle:

Step 1: a proton hits a ^{12}C nucleus (^{12}C is ordinary, common carbon, which has the atomic weight 12), fusing to ^{13}N (a radioactive isotope of nitrogen) and emitting gamma radiation;

Step 2: the radioactive ^{13}N decays to ^{13}C, emitting a positron (positive anti-electron) and a neutrino;

Step 3: ^{13}C collides with a proton, fusing to ^{14}N and emitting a gamma ray;

Step 4: ^{14}N collides with a proton, fusing to ^{15}O, a radioactive isotope of oxygen, and emitting a gamma ray;

Step 5: ^{15}O decays to ^{15}N, emitting a positron and a neutrino;[27]

Step 6: ^{15}N collides with a proton, fissioning to ^{12}C and an alpha particle.

This ends the cycle as it began, with a ^{12}C nucleus. The carbon nucleus thus acts merely as a catalyst and no carbon is actually used up. The end result is that four protons are fused into an alpha particle. The four nucleons (two protons and two neutrons) in an alpha particle being particularly tightly bound together, their fusion releases an especially large amount of energy.

The Sun's interior is also sufficiently hot for a second fusion cycle called the pp cycle, suggested by Hans Bethe, which consists of only three steps: in the first step, two protons fuse, forming a deuteron and emitting a positron and a neutrino; in the second step, the deuteron fuses with another proton to form ^{3}He, emitting a gamma ray; in the third step, two ^{3}He nuclei fuse to form an alpha particle, and two protons escape. Again, the end result is that four protons have been fused into an alpha particle, producing kinetic energy — heat — and gamma radiation as well as neutrinos. The effect of these sequences of events is to generate kinetic energy, and thus heat, as well as gamma

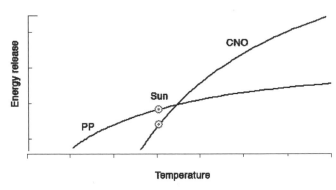

Figure 3.1. Graphs of the temperature dependences of the energy release caused by the pp and the CNO cycles.

radiation. The competition between the two cycles depends on the temperature, as depicted in Fig. 3.1.

Of the emissions, neither the positrons nor the gamma rays are able to escape from the interior of the Sun, as they all collide and interact with other particles. Only the neutrinos, which interact so weakly with anything that even the core of the Sun is transparent to them, manage to get out. As a result, the neutrinos generated in steps 2 and 5 of the CNO cycle and in the first step of the pp cycle are able to escape, some of them reaching the Earth. It took a long time before these neutrinos were actually detected. There was a further delay because of what came to be known as the *solar neutrino puzzle*: only about one third as many neutrinos were found as should have been. This puzzle was not solved until 2003 by means of a newly evolved theory according to which there are three kinds of neutrinos that convert into one another in an oscillatory fashion. These neutrino oscillations reduce the number of the kind of neutrinos emitted by the Sun that are detected on the Earth, thus explaining the puzzle. When all this was finally ironed out, the observed energy distributions of the solar neutrinos corresponded to just what they should be according to the Bethe-Weizsäcker CNO cycle and Bethe's pp cycle, confirming their theory that these sequences of nuclear reactions are indeed the main contributors to the heating of the Sun: about 1% of the heat and electromagnetic energy of the sun is generated by the CNO cycle, and the remaining 99% by the pp cycle.

Hans Albrecht Bethe, the physicist who contributed so much to our understanding of the functioning of the Sun and the stars, was born in 1906 in Strasbourg (which was German at the time) and educated at the Universities of Frankfurt and Munich, followed by postgraduate work at Cambridge University and with Enrico Fermi in Rome. After making fundamental contributions to solid-state physics in the late 1920s, he left Germany when the Nazis came to power, as his mother was Jewish. First moving to Manchester, England, followed by Bristol, he moved to the United States in 1935, where he joined the faculty at Cornell University, remaining there for the rest of his life, specializing and making basic contributions in nuclear physics as well as in quantum electrodynamics.

During the Second World War Bethe joined the Manhattan Project at Los Alamos as head of the theoretical division. Remaining active in nuclear political issues after the war, his consistent stance for peaceful resolutions gained him wide admiration, both as a physicist and an enlightened thinker. Honored with the Nobel Prize in physics in 1967 as well as many other awards and pursuing research, mostly in astrophysics, until the very end, Hans Bethe died in Ithaca, New York, in 2005 at the age of 98.

Energy in the Stars

Other stars are known to be considerably hotter than the Sun, so that other thermonuclear reactions come into play as well, such as the following.

The first, called the PPII chain or lithium burning, has lithium and beryllium as an intermediate:

Step 1: ^{3}He fuses with ^{4}He to form ^{7}Be (which has four protons) plus a gamma ray;

Step 2: ^{7}Be plus an electron forms ^{7}Li plus a neutrino (converting a proton plus an electron into a neutron and a neutrino;

this process is the converse of the beta-decay of the neutron, which will be discussed in Chapter 4);

Step 3: finally, ^{7}Li combines with a deuteron, i.e., ^{2}H, to produce two alpha particles, ^{4}He.

This process requires a temperature above 14 million degrees K, which is hotter than the core of the sun.

Another chain of reactions, called PPIII, begins with the beryllium produced in the first step of the PPII sequence described above and has it collide with a proton, capturing it to form ^{8}B, an unstable isotope of boron, plus a gamma ray. The ^{8}B nucleus then decays into the nucleus of ^{8}Be plus a positron and a neutrino, and the nucleus of ^{8}Be subsequently decays into two alpha particles. This chain requires even higher temperatures — above 23 million K — than the PPII chain.

However, the energy source of the enormous stellar explosions that astronomers occasionally observe in the form of a supernova is not nuclear. The evolutionary end of stars of about 0.07 to 10 solar masses is a star called a 'white dwarf,' so dense that the atomic electrons have no room to circulate in their orbits and atoms cease to exist. If such a white dwarf has a nearby companion, it may attract it, swallow it up and grow by accretion to a mass greater than 1.4 times that of the sun, called the Chandrasekhar limit after the great Indian-born American astrophysicist known to his friends as Chandra.[28] When that happens, or if its size is beyond the Chandrasekhar limit to start with, it has such strong internal gravitational pull that eventually it has to collapse into itself, producing a spectacular supernova and ends up as an enormously dense neutron star, a little more massive than the Sun but only a few miles in diameter.

The two sequences of thermonuclear reactions, the pp cycle and the CNO cycle described above, produce enough kinetic energy to keep the Sun hot for many billions of years. But there still remains the question of why it radiates so much energy in the form of visible light, electromagnetic waves of wavelengths that we perceive as the colors of the rainbow, as well as the ultraviolet radiation of shorter and infrared radiation of longer wavelengths. To answer this, we shall have to make an excursion into the quantum theory.

Notes

[21] The raised number to the left of the letter of the element indicates the atomic weight of the isotope. An isotope is one of several forms of the same element with different atomic weights. This means that the atoms of isotopes of a given element differ in the number of neutrons contained in their nuclei.

[22] The neutrons first produce ^{239}U, which beta-decays into neptunium, ^{239}Np, which in turn beta-decays into ^{239}Pu.

[23] The total size of a fission explosion, though enormous, is limited by the fact that if more than a critical mass is assembled in one piece, it will immediately tend to explode, as there are always a few stray neutrons around in the atmosphere. Chunks of ^{235}U therefore have to be handled with great care.

[24] Water made up of oxygen and deuterium, ^{2}H$_2$O, is called heavy water.

[25] That's also the reason why this nucleus is stable, even though the neutron is subject to beta decay, as discussed in Chapter 4. The decay products would have more rest energy than the deuteron. The same is true, *mutatis mutandis*, for all other stable nuclei in the periodic table.

[26] This is the absolute temperature scale, the letter K indicating degrees kelvin, which uses celsius degrees but starts with zero at $-273.16°$ Celsius. It is not customary to use ° to indicate degrees kelvin.

[27] Note that the well-verified radioactivity of ^{13}N and ^{15}O, the former with a half-life of a little under ten minutes and the latter of about two minutes, is unusual in that they emit positrons rather than electrons, as radioactive elements usually do.

[28] See *Physics Today*, December 2010, for a biographical portrait.

4

Energy in Quantum Mechanics

Quantum mechanics, the revolutionary new manner of describing the physical world, was initiated by Niels Bohr and Albert Einstein but molded into a coherent theory in the 1920s by Werner Heisenberg, Erwin Schrödinger, Max Born, and Paul Dirac. It transformed the way the energy of sub-microscopic systems was seen to behave. Here is the way it all began.

When Ernest Rutherford's experiments led to his discovery of the tiny nucleus at the center of atoms, with a positive electric charge and almost all the atomic mass concentrated there, he envisaged an atom as a miniature solar system, with the very light and negatively charged electrons circling about the nucleus like planets around the Sun, with gravity replaced by electrostatic attraction. The trouble was that according to Maxwell's laws of electromagnetism such a system could not last: the circulating electrons would have to continually emit radiation, thereby gradually losing their energy and crashing into the center. The basic idea of Rutherford's atomic model clearly made sense, given his astonishing experimental results that showed the atom to be almost entirely empty with a tiny nucleus at the center. However, it simply could not be reconciled with classical physics without drastic changes in its laws.

The necessary revision was introduced in 1913 when the young Danish physicist Niels Bohr made the revolutionary suggestion that the energies of the circling electrons had to be confined to specific, discrete levels. While there, they would not radiate, he decreed, Maxwell's laws notwithstanding. Any electron at a level higher than the one with the lowest energy (called the ground state) could make a sudden leap to a lower level, the energy difference between the two being emitted in the form of electromagnetic radiation. The frequency of the emitted radiation he postulated to be equal to this energy difference divided by a number that Max Planck had introduced a few years earlier, Planck's constant, for the purpose of accounting for the distribution of the radiation emitted by a hot black body.[29] All of this was purely *ad hoc*, with no underlying coherent theory — it was nothing but Bohr's brilliant intuition to account for experimental evidence.

Niels Bohr, at the time working as an assistant in Rutherford's laboratory in Manchester, had been born in 1885 in Copenhagen, his father a professor of physiology at the University of Copenhagen. After earning his doctorate there with a dissertation on the behavior of electrons in metals, he left for Cambridge for post-doctoral work. Not finding the atmosphere at Cambridge congenial, he moved on in 1912 to Manchester, where Rutherford had just discovered the atomic nucleus. He stayed for four years — interrupted by returning to Copenhagen to get married — until he was appointed Professor of Physics at Copenhagen University. Meanwhile he was becoming famous, as his model of the atom, its revolutionary features violating classical theory notwithstanding, seemed to be very successful in correctly predicting and accounting for new experimental data. His model looked as though it might actually serve to explain the periodic table of the elements! The Danish government, proud of its famous citizen, set up an Institute of Theoretical Physics and made him its director. A year later he was awarded the Nobel Prize in physics.

He had six sons; the eldest died tragically at the age of sixteen in a sailing accident, his father watching but unable to help; another son, Aage, became a Nobel-prize winning physicist in his own right. Niels Bohr remained at his Institute, to which physicists flocked from all over the world, until his death in 1962, the grand old man of the quantum theory, defending it against many criticisms, including those by the physicist he admired most, Albert Einstein. His philosophizing about its strange implications became very influential, though not all physicists accepted his sometimes impenetrable pronouncements.

The scaffolding of the architecture of the full theory, called *quantum mechanics*, was designed by Heisenberg and Max Born in Göttingen, Germany, Schrödinger, an Austrian in Zürich, Switzerland, and Dirac in Cambridge, England. Incorporating the ideas of Planck, Einstein, and Bohr, it resembles the structure of Newton's mechanics and Maxwell's electromagnetism. However, it differs drastically from both by replacing any attempt at directly describing the real world and its behavior with the mathematical construction of a remote abstraction. Whereas the classical theories precisely predict the future motions of particles on the basis of their present state, quantum mechanics merely predicts probabilities. The consequences of this radically new theory are far-reaching, but contrary to some popular stories, they are not confined to accounting for nature as though she resembled a giant crapshoot. When it comes to energy, quantum mechanics, together with its generalization of Maxwell's electrodynamics (called quantum electrodynamics, or QED for short), is able to make very precise predictions that differ from those of classical mechanics. These pedictions are extremely well borne out by experimental observations, confirming the theory.

One of the applications of quantum mechanics relates to our discussion of radioactivity in the last chapter, specifically alpha radioactivity, in which an atomic nucleus emits a helium nucleus. We have to begin with the fact, mentioned earlier, that the four nucleons in an alpha particle are particularly strongly bound together, so that inside certain heavy nuclei containing many protons and neutrons they form clumps that can be regarded as particles all by themselves. These alpha particles are subject to two strong forces there: the electrostatic repulsive force because they are positively charged as are all the protons around them, and the strongly attractive short-range nuclear force among all nucleons. The addition of these two forces, described in terms of potential energy, results in a barrier of a certain calculable height and width, rounded at the top, surrounding the interior of the heavy nucleus. Their kinetic energy being lower than the height of the barrier, they cannot escape, and the heavy nucleus is stable — when looked at classically. However, the Russian-born American physicist George Gamow argued that, according to quantum mechanics, even particles with insufficient kinetic energy to surmount a barrier of potential energy have a certain probability of 'tunneling' through it, and he was able to calculate the escape probability per unit time. The higher their kinetic energy — the faster they move — inside the nucleus, the closer that energy is to the top of the barrier, and the larger their escape probability. Therefore quantum mechanics allowed him to calculate, for a given heavy nucleus, how long it will take until there is a 50/50 chance for one of the alpha particles to have escaped: this is the half-life of a nucleus subject to alpha radioactivity, and the higher the energy of a given alpha-decay, the shorter its half-life. The probabilistic new theory thus accounts very well for the mysterious fact that the time of radioactive decay of a given nucleus cannot be predicted, but the time after which half a large collection of the same kind of alpha-radioactive nuclei have decayed can be accurately predicted. This theory also agrees with the observed fact that the individual alpha particles emitted by a chunk of alpha-radioactive material appear quite irregularly, seemingly at random intervals. The explanation of beta radioactivity took considerably longer, and we shall return to it shortly.

Discrete Energy Levels

Bohr's revolutionary insight that initiated the quantum theory of atoms was retained in the fully developed theory. The energy of a planet circling the sun can have any numerical value below the one it has when at rest infinitely far away where its potential energy vanishes — in other words, any negative value — which makes it unable to escape. Quantum mechanics, on the other hand, says that an electron in an atom — naively pictured as circling the nucleus like a planet, the way Rutherford had it — can have only very specific, discrete values that can be quite precisely calculated by the new rules. What is more, in an atom with more than one electron, no more than two (one for each value of its two 'spin' directions) can occupy each energy level.[30] The outermost electrons of atoms are the ones involved in forming the chemical bonds between these atoms when building molecules and in all chemical reactions. As a result, the quantum-mechanical energy rules produce a rational explanation of the periodic table of the elements, the backbone of all of chemistry. This table had been postulated at the end of the nineteenth century by Mendeleyev entirely *ad hoc* on the basis of the elements' chemical properties.

The same peculiarity, that is, that energy is allowed to take on only certain specific discrete values, holds for any other physical system of particles that remains bound together, not just for atoms. For example, the motion of a simple pendulum, which classically is described as a back-and-forth oscillation at a fixed frequency, swinging at a wide continuous range of amplitudes or energies, has quantum mechanically an infinite set of discrete energies assigned to it. This set of energies has two particular characteristics that turned out to be important in later developments: the steps in the ladder of allowed energies are all equally spaced, and even the ground state, when the pendulum is classically standing still, has a certain 'zero-point energy,' which is a fixed multiple of its frequency.

The first of these features can be regarded as an explanation of the existence of Einstein's photons.[31] The second turned out to produce one of the strange results of QED, the quantum version of Maxwell's electrodynamics, because it employs a model that decomposes all

51

electromagnetic radiation into a sum of infinitely many mathematical pendulums of different frequencies. If all of these have a fixed zero-point energy, then even the 'empty' vacuum contains energy! What is more, this infinite sum of fixed zero-point energies adds up to an infinite amount, an obstacle that took considerable ingenuity to overcome. However, when this difficulty was finally surmounted independently by two Americans, Julian Schwinger and Richard Feynman, and the Japanese physicist Sin-itiro Tomonaga, its results turned out to agree with experimental data on the fine details of atomic energy levels as well as the magnetism of the electron to better than one part in a billion. Even though the 'renormalization method' employed to get rid of its infinities was considered ugly by many physicists, based on its predictive power QED may be regarded as the most successful physical theory ever invented.

In the course of the twentieth century, physicists devoted much time and effort to the construction of quantum field theories, that is, quantum theories that not only generalized Maxwell's electrodynamics but other theories of a similar nature. Their aim was to account for the existence of a very large variety of elementary particles that had been produced by collisions between fast-moving particles. High-energy physicists — a newly coined specialty — had constructed larger and larger machines that accelerated charged particles such as protons or electrons to higher and higher speeds.[32] The reason why these high energies were required is that in order to produce a particle of mass m the collision has to have a minimal kinetic energy equal to mc^2, according to Einstein's equation. Therefore looking for new particles of larger and larger masses required collisions of higher and higher energies.[33]

The aim of the new quantum field theories was to account for these newly discovered particles by means of the precise, discrete quantum energy rules. The principal tool here was again Einstein's $E = mc^2$: if the quantum rules lead to discrete energies, then Einstein's relation decrees the masses of new particles. These efforts turned out to be very successful, despite the fact that the postulated field equations were much too complicated to be solved exactly. What helped enormously was that simply postulating that the equations obeyed certain — sometimes very abstract — symmetry rules was sufficient to draw important conclusions about the allowed energies — and hence masses — and other properties of the implied particles.[34] From the perspective of this book, the notable point is that physical theory leads from quantum-mechanical energy rules to the explanation of the existence of all the particles of the universe.

We should now return more specifically to atoms, but before we do we have to discuss another strange rule of quantum mechanics.

While Heisenberg's uncertainty relation, which prevents us from knowing simultaneously the position and momentum of a particle with unlimited precision, is well known, when quantum mechanics is discussed an analogous relation between energy and time is rarely mentioned. In a sense it states that the product of the precision with which the energy of a system is determined and the length of time during which that system remains in the state of that energy has to be greater than Planck's constant h. This may be interpreted to mean that the sacred law of conservation of energy can be violated by a certain amount A for a short period of time T, provided that T is no larger than h/A. Here is an example pertaining to QED.

Electromagnetic radiation, according to QED, can create particles, provided all the relevant conservation laws are satisfied. For example, a photon of energy E can produce an electron-positron pair if E is greater than $2mc^2$, where m is the mass of the electron (which is equal to that of the positron) — it has to be a pair so that the total electric charge remains zero. However, according to the relaxation of the law of energy conservation just mentioned, a photon with less energy can also create a 'virtual' pair for a brief length of time less than $h/2mc^2$. (Note that the law of conservation of charge must *not* be violated!) The result of such weird fluctuations giving rise to the 'existence' of charged particles in vacuum is that even the vacuum has certain electromagnetic properties — such as polarization — that resemble those of matter. Another application of the 'energy-time uncertainty relation' will be given shortly.

The discrete nature of atomic energy levels turned out to be relevant also for beta-radioactivity. Quite surprisingly it was discovered that the neutron, one of the fundamental building blocks of the atomic nucleus, is unstable. With a half-life of about ten minutes, the free neutron, slightly heavier than the proton, decays into a proton, an electron, and a neutrino,[35] all with a relatively small amount of kinetic energy, as the mass excess of the neutron over the sum of the masses of the proton and the electron is quite small. You might therefore expect every atom heavier than hydrogen — the only one whose nucleus contains no neutrons — to be radioactive. What saves the world from this calamity is the fact that the constituents of most nuclei are so tightly bound together that the rest energy of these nuclei is less than the sum of the energies of all the particles emerging if any of the neutrons were to decay. These nuclei are therefore prevented from radioactive decay by energy conservation, just as the deuteron is (see Chapter 3).

Of course, there remains the question of the decay of the free neutron. The explanation of that was considerably more difficult and took much longer, as it required the invention of an entirely new field theory. Fermi tried to construct a theory modeled after the enormously successful QED, but it failed. The principal difficulty was the fact that in quantum electrodynamics the fundamental interaction involved just two particles, the electron and the photon, whereas in beta decay it took three particles to describe the basic weak interaction: the nucleon (either a neutron or a proton), the electron, and the neutrino. After a number of years, three physicists independently developed a successful theory: the Americans Sheldon Glashow and Steven Weinberg at Harvard, and the Pakistani Abdus Salam at Imperial College, London. A generalization of electromagnetism, it came to be called the 'electroweak' theory.

However, it would take us beyond the reach of this book to explain it in detail.[36]

Quantum Jumps and Spectra

Now, what about Bohr's 'quantum jumps'? Just as Bohr had postulated in his original model, when an electron finds itself on an energy level above an empty place at a lower level, it can jump down, emitting electromagnetic radiation of a frequency equal to the energy difference divided by Planck's constant. Every element therefore has a characteristic set of frequencies of radiation it can emit, including light as well as possibly infrared and ultraviolet radiation — its spectrum. The probability per unit time for such a jump to happen can in principle be calculated.[37] How could an electron land up there at an elevated spot? That could happen if the atom got 'excited' by absorbing a photon of just the right energy, or by colliding with another atom of sufficient kinetic energy. When a given element is heated, the energetic collisions excite its atoms, i.e., they kick some of their electrons to higher levels and, as these electrons descend, they will emit the radiation 'line spectrum' (the photograph of the emitted light, when passed through a spectrograph, shows one vertical line for each emitted frequency; see Fig. 4.1 for an example) characteristic of that element, a fact with which chemists had long been familiar without understanding its origin. It is the reason why sodium light is yellow — throw a grain of salt into a candle flame, and you will see — while the light emitted by mercury vapor has a bluish tinge. The spectacular display of the *aurora borealis* (northern

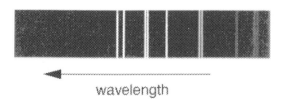

wavelength

Figure 4.1. Image of the line spectrum of mercury.

55

Figure 4.2. The *aurora borealis* visible in Alaska

lights) visible at night in the northern latitudes is the result of such excitations of atoms in the upper air (the thermosphere) by collisions with charged particles coming in from outer space and the sun, which had been concentrated near the pole by the Earth's magnetic field. The excited atoms subsequently decay, emitting their characteristic line spectra.

Now here we come again to the 'energy-time uncertainty relation' mentioned earlier. An energy level of an atom that is forever stable, like its ground state, can be determined — in principle — with unlimited precision. However, if a level is unstable, like that of an atomic electron when a lower level is unoccupied, so that the electron has a certain probability of descending with the emission of a photon, then its energy cannot be precisely determined; it necessarily has a certain 'breadth' given by Planck's constant divided by that emission probability. As a result, what we usually call 'line spectra' are not really completely sharp lines, but each has a certain specific width.

However, this lack of sharpness of spectral lines is small enough to be negligible for most purposes.

So we are now finally in a position to explain why we receive the life-giving energy of the Sun in the form of light of the colors that Isaac Newton saw when he decomposed sunlight for the first time by means of a prism, and more. In addition to these visible colors, there are the invisible ultraviolet (high frequency and hence high-energy, penetrating 'hard' photons) and the warming infrared ('soft' photons, low-frequency radiation which is able to make molecules vibrate, producing heat but unable to penetrate far into matter) "colors".

The visible surface of the Sun consists of a mantle of very hot gas (though not hot enough to sustain thermonuclear reactions) called the photosphere, made up of many elements. Their atoms get excited by collisions and each element subsequently emits its characteristic spectrum. This is what we receive as sunshine.

Sunlight subjected to spectral analysis during the solar eclipse in 1868, when nothing but the corona — further out than the

Figure 4.3. An image of the corona of the Sun taken during a solar eclipse.

photosphere and usually invisible — could be seen, was found to contain, in addition to many spectra of familiar elements, one spectrum never seen on Earth before. The corresponding element was named *helium* after Helios, the ancient Greek god of the Sun. It took twenty seven years until it was found on the Earth as well, nicely filling an empty slot for a noble gas in the periodic table of the elements. As we have seen, helium, element number 2 in the periodic table with an atomic weight of 4, plays a variety of important roles in physical processes.

The most important aspect of the energy we receive from the Sun is, of course, that half the turning Earth bathes in it regularly, every day. But, in addition, some of it has also been stored, and we have learned to retrieve this highly valued treasure. To that, and other methods of energy storage as well as transmission, we shall now turn in the next chapter.

Notes

[29] Seven years before Bohr's model of the atom, Einstein had already introduced the idea that all electromagnetic radiation was made up of 'quanta,' whose energy was equal to the frequency of the radiation multiplied by Planck's constant. Bohr had not yet been convinced of that, though eventually, of course, he would be.

[30] This is called the 'exclusion principle.'

[31] For more details on this, see my book *Galileo's Pendulum: From the Rhythm of Time to the Making of Matter.*

[32] These energies were measured in units of eV, or electron volts, i.e. the energy an electron attains after being accelerated by a voltage difference of 1 volt. Then larger units were introduced: MeV, a million eV, and GeV, a billion eV.

[33] Masses were now measured in the new energy units too; the mass of the electron is about 0.5 MeV, that of the proton approximately 1 GeV.

34 For more details, see, for example, my book *From Clockwork to Crapshoot: A History of Physics*, as well as *Galileo's Pendulum*, op.cit.

35 More accurately, anti-neutrino.

36 For the history of its development, see for example my book, *From Clockwork to Crapshoot*.

37 Indeed, Einstein had already found a correct way to calculate this probability before the set of rules of the new theory were promulgated.

5

Storing and Transporting Energy

Long-term Energy Storage

As we have seen in Chapter 2, a large part of the energy we continually receive from the Sun in the form of light is converted into chemical energy. The chlorophyll contained in all green plants produces enormous amounts of carbohydrates by the process of photosynthesis. What happened to all these combustible compounds generated in the course of the Earth's long history? Much of that energy was stored by being turned into what we now call fossil fuels.

It began at a time, hundreds of millions of years ago, before mammals existed and even before the dinosaurs started roaming the Earth. The atmosphere was hot and much of the ground was swampy, with huge trees and other large plants growing on it. Their detritus fell to the ground the way leaves do today. Dead trees, other dead plants, and their greenery simply accumulated, festering and decaying for millions of years at the bottom of swamps, eventually covered by thick layers of dirt and water, all exerting pressure on the mass underneath. As the climate and the shape of the Earth's surface changed and the ground shifted, further decay halted as no oxygen could reach the tightly covered biomass. This process continued for millions of years, some of it in the course of time ending up deep underground and some closer to the surface, by changing geology, all turned black and almost as hard as stone: it is now coal. (The same kind of biological detritus of more recent, even contemporary origin is also convertible into usable energy; it is then referred to simply as biomass.)

The existence and usefulness of coal for home heating purposes and metal working has been recognized since antiquity, as described by the Greek geologist Theophrastus, a student of Aristotle's, in the fourth century BCE. During their occupation of Britain the Romans mined for coal there as well. During the Ming Dynasty, the Chinese dug for coal in open pits and shallow tunnels. By that time, in the late sixteenth and early seventeenth century, mining had begun in the English coalfields and coal was used in households as well as in the manufacture of bricks and earthenware. In the Americas, the Aztecs and other native Americans utilized coal both for heating and for baking clay pottery long before the continent's discovery by Europeans. In the early seventeenth century their conquerors began to find coal useful as well, and eventually it became the main source of energy for the heat that propelled the industrial revolution. The chimneys and smoke stacks spewing forth gray smoke from its burning for home heating in the winter and for driving the machines of factories all year around became the recognizable dark landmark of the cities of the Western, and eventually of all the 'developed' and much of the 'developing', world. Well into the twentieth century, London was famous for its constant 'mist,' reducing visibility sometimes to a few yards. The mist was nothing but dense smog that blackened its entire architecture until it was cleaned up in the 1950s and the use of coal burning reduced. After China began its great industrialization, the city of Beijing went through a similar experience that is still not resolved.

Now, as we saw in Chapter 2, green plants are not the only users of photosynthesis; so are blue-green algae and many bacteria, as well as general zooplankton. During the age of the dinosaurs, the water near the ocean surface was warmer than today, the upper regions teeming with microscopic life. As the bacteria, algae, and diatoms died, they sank to the dark bottom, eventually forming a thick sludge that was

Figure 5.1.   A typical operating oil well on land.

heated by the warm Earth, part of it trapped under layers of rock and sediment as the geology was altered. Under contant pressure, some stayed liquid, slowly turning into oil, while some turned into gas. As a result of the shifting geology caused by sliding tectonic plates on the planet's surface, the distribution of the oceans changed and large areas once covered by water turned into dry land. Many oil and natural gas deposits, all originating long ago under water, can thus be found now in deserts and prairies far from the ocean as well as offshore.

Consisting of many different hydrocarbons with a great variety of weights and volatility, from tar-like asphalts to gas, the composition of these petroleum deposits varies from one location to another, some of them mixed with sand. Asphalt deposits found near the surface were reported to have been used as early as some four millennia ago in the construction of the towers of Babylon. During the Han dynasty in China two thousand years ago, petroleum seems to have been used as an incendiary weapon.

In more modern times, after the Polish petroleum pioneer Ignacy Łukasiewicz invented a process to distill kerosene from petroleum in the nineteenth century, kerosene lamps became important for lighting, taking the place of lamps fueled by whale oil. In fact, both the world's first oil well and the first oil refinery were built in Poland at that time.

It was, of course, the invention of the internal combustion engine, with the subsequent development of the automobile, that led to an explosive thirst for oil, and especially gasoline, in the United States and Europe. Gasoline, just like kerosene, is one component of crude oil, so refineries were urgently needed and built to distill petroleum.[38]

Coal and oil, as we have seen, are nothing but large amounts of the Sun's energy reaching the Earth held in long-term storage, and both originate from the process of photosynthesis by living organisms. Neither kind of fossil fuel can therefore exist on the Moon or on any planet that never developed carbon-based life.

Now let us turn to the storage of energy for shorter times.

Short-term Energy Storage

Flywheels

The simplest device for the short-term storage of kinetic energy is the flywheel. It consists simply of a heavy wheel with a large moment of inertia which, once set in motion, retains its kinetic energy until friction brings it to halt, at which point all the kinetic energy has been converted into heat. Early gasoline and diesel engines, which by their design produce rotational kinetic energy in short bursts rather than continuously, were connected to flywheels so as to even out their delivery of mechanical energy. Later designs made this unnecessary by combining the combustions in several cylinders in such a way that they delivered their thrusts more evenly. That's why the more cylinders a car engine has, the more smoothly it runs. Heavy vehicles, such as big locomotives and ocean liners driven by diesel engines, have enough inertia on their own to make the smoothing action of flywheels unnecessary.

Automobile companies have begun to use what they call the Kinetic Energy Recovery System (KERS), which uses a flywheel for braking. Instead of wasting the lost kinetic energy by converting it into heat, as in conventional brakes, KERS stores it temporarily in a flywheel, to be used later, helping to accelerate. The Volvo Car Corporation has recently introduced it, and Honda, Peugeot, as well as Mercedes are testing KERS and are planning to use the system.

Pumped Storage

Whereas flywheels temporarily store energy in its rotational kinetic form, pumped storage saves it in the form of gravitational potential energy. This method is used by some hydroelectric power plants, whose energy source, such as a waterfall or the water flow from a dam, runs at a naturally or artificially limited rate, to even out periods of low use of electric power and periods of high use. The plant utilizes a large separate water reservoir at a higher elevation — either naturally existing or specially created — and during off-peak times it uses some of its unused power to pump water to it, storing it there. During times of peak demands of electricity, it then drives additional turbines by opening water flow from this storage reservoir. In contrast to flywheel KERS, the pumped storage does not prevent the loss of useful energy to heat; it merely saves the fuelling energy source of the plant for use during another time period.

Batteries

The most important practical short-term energy-storage device is the electrical battery.[39]

The battery is really a device for converting electrical into chemical energy, storing it in that form, and reconverting it back into electrical energy. It consists of two metal 'electrodes' inserted in a (usually liquid) 'electrolyte.' In a typical car battery the electrolyte consists of sulfuric acid, with the chemical formula H_2SO_4, dissolved in water, where its molecules split into positively charged hydrogen ions and negatively charged SO_4 ions.[40] One of the electrodes is usually made of lead, the other of lead dioxide, PbO_2. When these are inserted in the electrolyte, the negative ions are attracted by the lead dioxide and remain there until there are so many coating it that their negative charge is sufficiently strong to repel further SO_4 ions, while the positive ions wander to the lead electrode until it is saturated as well. The negatively charged electrode is called the *cathode*, and the positively charged, the *anode*. In other batteries, the electrodes may be made of different metals, and the electrolyte of other chemicals, but all with similar chemical properties; see Fig. 5.2.

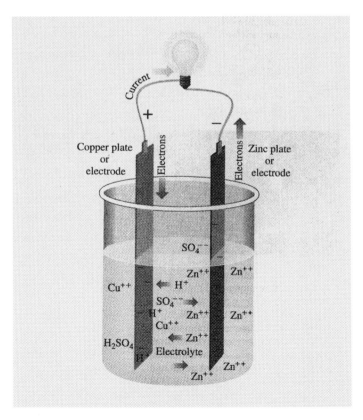

Figure 5.2. A wet cell battery with electrodes plated with copper and zinc.

The battery is now charged, and there is a voltage difference between the two terminals. When these are connected by a wire to an electric device such as a light bulb or a motor, current — electrons, not SO_4 ions — will flow through the wire from the cathode to the anode. This will neutralize the charges accumulated at both, allowing more negative ions to be drawn to the cathode and more positive ones to the anode, until all the ions in the electrolyte are converted into neutral molecules and the process ends; the battery is discharged. If a current is made to flow in the opposite direction, the electrodes will act to give again a negative charge to the SO_4 molecules and a positive charge to the hydrogen molecules, thus recharging the battery and again storing energy in it in chemical form.

Any given battery cell produces a specific voltage difference between its terminals. In order to produce larger voltages, several cells have to be connected by wire 'in series,' that is, each cell's cathode to the next cell's anode, etc.

This extremely useful device was invented in 1800 by the Italian physicist Alessandro Volta[41] after the anatomist Luigi Galvani had noticed that the dissected leg of a dead frog twitched when he touched them with two different metals. Experimenting for years, he managed to produce 'animal electricity' by connecting a frog's leg to two different metals, multiplying the effect by connecting them 'in series,' and finding that the result made another frog's leg twitch again.[42]

Useful as the battery is for energy storage, say for the purpose of starting and propelling cars, its usefulness is limited by its bulkiness and weight when a large amount of energy is required, and there has not been as much advance in increasing storage capacity and weight reduction as needed. The invention and development of a better, lighter large-capacity battery is one of the most urgent needs in the energy field, and many scientific laboratories are at work on its solution.[43] However, electric cars powered by batteries were already proposed in the nineteenth century, and modern versions, of course, exist as well, made by Chevrolet as well as Nissan.

Figure 5.3. A battery-operated car designed in 1895 by Thomas Edison.

Non-rechargeable batteries differ from most rechargeable ones by using dry rather than liquid electrolytes. In the case of alkaline batteries it is potassium hydroxide, and the anode is composed of zinc powder, while the cathode is of manganese dioxide. There has been much progress in the miniaturization of such batteries, for a great variety of uses. However, as yet there is none capable of storing large quantities of energy.

Liquid Hydrogen

Liquid hydrogen by itself can also serve as a medium for short-term energy storage. When an electric current flows through water, it causes the H_2O molecules to be be decomposed into positive hydrogen ions, collecting at the cathode (the negative end of the wire entering the water), and negative oxygen ions at the positive anode. The hydrogen gas can then be separated and used as is or liquefied by lowering its temperature and raising its pressure. As soon

Figure 5.4. A modern three-wheeled electric car made by Aptera Motors.

as such hydrogen comes in contact with oxygen, it is easily ignited to produce a strong exothermal chemical combustion. In effect, the electrical energy spent in the electrolysis of water has been stored in the hydrogen, and it is recovered in the form of heat when the hydrogen is burned, producing water. However, hydrogen may also be used to produce electrical energy by a fuel cell.

Fuel Cells

A hydrogen fuel cell consists of two chambers, one filled with liquid hydrogen, connected to a hydrogen reservoir, and a catalyst (such as very fine platinum powder) that ionizes the hydrogen atoms, the other filled with oxygen. The two chambers are separated by a region filled with an electrolyte that allows protons — hydrogen ions — to pass through, but not electrons. The result is that the hydrogen ions reaching the first chamber combine (assisted by another catalyst such as nickel) with the oxygen within to form water. Because the electric neutralization of these ions require electrons, an electrode inserted in this chamber becomes an anode, sucking in electrons, so to speak, when connected to a wire. Similarly, an electrode inserted in the first chamber turns into a cathode, pushing out surplus electrons left over from the loss of hydrogen ions. That's how the fuel cell turns the energy temporarily stored in hydrogen into an electric voltage (usually about 0.6 Volts).

It should be clearly understood that the use of liquid hydrogen to store and transport energy (for propelling automobiles, say) is fundamentally different from the use of gasoline. The latter, having been simply extracted from the ground and refined, is for us a *source* of energy, whereas the former is merely a vehicle for energy transport, since it requires as much energy (or more, after some inevitable loss) to generate the hydrogen as is produced when the fuel cell turns it into electric energy. Liquid hydrogen is in no sense a substitute for gasoline.

Energy Transport

By far the simplest and most efficient form in which energy can be transported over large distances is electromagnetic, and the overwhelmingly largest amount of it that we witness every day is, of course, the Sun's radiation coming to us through space. Hugely beneficial as it is, this particular mode of transport, however, is not a useful model for us to emulate, as it is enormously wasteful. After all, of all the energy radiated by the Sun, our planet receives only a tiny fraction. In order to transport electromagnetic energy from one point on Earth to another distant point we cannot simply do it by means of large searchlights and the needed mirrors to overcome the curvature of the Earth. The laws of optics do not allow the beam to be sufficiently focused to prevent great losses; any such endeavor would thus be extremely inefficient.

In principle, there is, however, a closely related way that prevents the defocusing of ordinary light: the laser beam. The working of a laser[44] is based on an ingenious early suggestion of Einstein's.

Shortly after Bohr had introduced his revolutionary idea of rendering Rutherford's model of the atom stable — electrons circulating about the nucleus in discrete orbits and radiating only when jumping from one level to a lower one — Einstein suggested a formula for calculating the probabilities of these quantum jumps that correctly anticipated just what the fully constructed quantum mechanics later predicted. At the same time, he also suggested that when an atom in an excited state — a state in which one of its electrons was in a position to jump to a vacant lower energy level — was struck by an electromagnetic wave of just the right frequency corresponding to the energy of this jump, it was more likely to execute the leap, and he produced a formula for the probability of such stimulated emission of radiation as well. It took many years before this fruitful idea was actually verified and utilized in the construction of the laser.

What distinguishes a laser beam from the beam produced by an ordinary source of light, such as a flashlight or the Sun, is that the light in the laser beam is *coherent.* This means that the light waves in it are all moving in step. If the light waves in an ordinary monochromatic beam can be compared to a group of strolling pedestrians, those in a laser beam are comparable to a column of soldiers marching in step. As a result of this coherence, such a beam can be focused very much more sharply, and it will retain this focus for long distances. Handheld laser pointers used by lecturers produce tiny bright points of light on projected images, and powerful laser beams have been used to produce bright spots bounced back by reflectors left by our atronauts on the Moon. This enabled astronomers to measure its momentary distance from Earth's surface to within inches.

Can laser beams, then, be used to transport electromagnetic energy over large distances on the Earth, with losses caused only by a certain amount of absorption in the atmosphere? If we should ever find energy sources on the Moon or on a planet such as Mars, this method of energy transport to the Earth would be the only practical one. Unfortunately no one has succeeded yet in constructing lasers powerful enough to make this idea practicable.

The way electricity is in fact transported over large distances currently is, of course, via high-voltage transmission lines. Why high-voltage? As we discussed in Chapter 2, when a current flows through a wire, it produces heat. When the purpose of the current is simply to transport energy, this heat is a loss and we want to minimize it. Now, the power loss to heat is given by the product of the square of the current I multiplied by the resistance[45] R of the wire, I^2R. In order to make the loss as small as possible, we want to make the current as small as possible.[46] But the power transported, as we saw in Chapter 2, is given by the product of the current multiplied by the voltage V. Therefore for a given amount of power, the current is minimized by making the voltage as high as possible: ergo the need for high voltage. In practice this means hundreds of thousands of volts. But that is not the kind of voltage we want for everyday usage in households; it would be much too dangerous. So we must be able to transform electricity from one voltage to another without much loss, and that is possible only for alternating current, not for direct current.[47]

The means to do that is called a *transformer*; it consists of a magnetizable iron core with two wires wound around it. If an alternating current is sent through one of the wires, it magnetizes the core by the effect discovered by Faraday (see Chapter 2), and the time-varying magnetic field so created produces a current in the second wire by Ampère's effect (again, see Chapter 2). The magnetization of the iron core being proportional to the number of windings of the first wire, as well as to the current in it, and the current in the second wire being proportional to its number of windings as well as to the magnetization of the core, the ratio of the voltages in the two wires is equal to the ratio of the number of their windings. The result is therefore a device that allows the transformation of an alternating current of one voltage into one of a different voltage. The high voltage advantageous for long-distance transport of electricity can thus be changed to a lower, less dangerous voltage for everyday usage.

Their inevitable losses notwithstanding, high-voltage transmission lines of alternating current are at present the most practical and efficient means of transferring energy from one point on Earth to another at large distances. There are other conveyances, such as oil and gas pipelines, tankers, and coal trains that may sometimes be preferable for a variety of reasons, but they tend to be much less efficient.[48] A hundred years from now transmission lines may well be replaced by high-energy laser beams (or maser beams, their equivalents for microwaves rather than light). As Niels Bohr is sometimes questionably quoted, 'prediction is difficult, especially of the future.'

We have thus reached the end of our discussion of storing and transporting energy, and thereby the end of relating the science of energy as concerns the Sun and the Earth. You may wonder at this point, where does all the energy in the universe come from? That cannot really be answered by physics since a fundamental law dictates that energy is and always has been conserved. Where it 'came from' may therefore be regarded as a philosophical rather than a scientific question. What was the form of energy at the beginning of the universe — if there was a beginning — and how did it change in the course of time? Those are more meaningful questions from a scientific point of view, and we shall turn to them in the next chapter.

Notes

[38] The process of extracting the various components of the crude oil as it comes from the Earth is based on their different boiling points, allowing fractional distillation to separate them.

[39] I am here concerned only with rechargeable ones, such as car batteries.

[40] This means that the two hydrogen atoms each lack one electron and the SO_4 quasi-molecules, called radicals, have two extra ones.

41 In his honor we use the names "voltage" for electrical potential differences, measured in units of "volts".

42 This shows why it is productive for scientists in different fields to talk to and learn from one another.

43 A recent book on the use of batteries, especially on cars: *Bottled Lightning: Superbatteries, Electric Cars, and the New Lithium Economy* by Seth Fletcher; see also the Appendix.

44 Laser is an acronym for *light amplification by stimulated emission of radiation.*

45 This resistance in turn is proportional to the length of the wire and inversely proportional to its cross section and depends on the material it is made of.

46 It would also be very advantageous to reduce the resistance of the wire to zero by cooling it sufficiently to make it superconducting; however, this is not yet technologically feasible.

47 Much of the credit for the universal use of alternating current for the transport of electric power belongs to the Serbian-American engineer and inventor Nicola Tesla, who prevailed over Thomas Edison's strong push for direct current.

48 Efficiency in this case is meant in a less technical sense and includes convenience and practicality.

6

Energy in the Universe

Ever since Einstein invented it, his general theory of relativity has been the framework for understanding how the universe evolved. And thereby hangs an intriguing little tale. Right after its announcement, his theory was used to construct a new mathematical model of the universe, and it was found that if his theory was correct, the cosmos could not be static: it had to be expanding or possibly contracting. That Einstein's theory implied such a development was first demonstrated by a Belgian Jesuit priest and cosmologist by the name of George Edouard Lemaître.

Born in 1894 in Charleroi, Belgium, Lemaître served as a decorated artillery officer in the Belgian army during the First World War and afterwards studied physics and mathematics while preparing for the priesthood. Ordained a Jesuit priest, he became a graduate student in astronomy at the University of Cambridge, learning about cosmology and stellar astronomy, followed by a year with the American astronomer Harlow Shapley at the Harvard College Observatory as well as at the Massachusetts Institute of Technology. On his return to Belgium he became a lecturer, and later a regular Professor at the Catholic University of Louvain, where he prepared a report, based on Einstein's theory of general relativity, with his new idea of an expanding universe (including Hubble's law two years before Hubble) that began with Einstein's original notion of a static version. Not long afterwards he came up with the idea that the universe expanded from an initial point, the 'Primeval Atom' or 'the Cosmic Egg exploding at the moment of creation.' Lemaître became quite famous, won a number of high honors and prizes, and was made a Monsignor by Pope John XXIII. He died in 1966 in Louvain, Belgium.

An expanding universe, however, was not what astronomers seemed to have observed. Working at the Mount Wilson Observatory in Pasadena, California in the early twentieth century, Harlow Shapley came to the astonishing conclusion that the universe was vastly larger than had previously been thought. How did he discover this?

Astronomers had always been stymied in all their attempts to measure the distances to the stars. Distances to the planets could be deduced from parallax measurements, a trigonometric triangulation using the variation in angles at which a given planet's position was seen during the Earth's motion.[49] But the stars and galaxies were much too far away for this method to work. Shapley instead used the known variable stars as distance markers. The brightness of these stars varies regularly, each with its own specific period of variation. This period was known to be correlated with the star's intrinsic brightness. Therefore, deducing the intrinsic brightness of such a star from its period and observing its brightness as seen through the telescope allowed him to calculate its distance, since the ratio of the observed to the intrinsic brightness should decrease as the square of the distance. These variable stars could thus serve as approximate distance markers for groups of stars around them and of galaxies to which they belonged. So far as astronomers could tell at the time, the conglomeration made up of all these stars and galaxies was huge, but it had a fixed size; it seemed to be static.

So Einstein had to do something about his general theory of relativity. Well, in order to assure that his equations did not lead to an expanding universe and contradict the observational evidence, he slightly modified them by introducing what came to be known as the *cosmological constant*. However, it did not take long until the astronomer Edwin Hubble discovered that the universe was not only enormously large, it was actually expanding. Einstein, of course, quickly removed the artifically introduced ugly constant from his equations, later regretfully saying that its introduction had been the biggest blunder of his life. Had he stood by his original theory he could have taken credit for predicting the finally observed expansion of the universe.

How did Hubble discover the expansion? When the spectra of light emitted by individual stars were examined in detail by astronomers, they found that these fingerprint-like identifiers of their constituent chemical elements emitted by them (see Chapter 4) showed that the stars consisted of the same elements known on earth. However, many of the individual spectra observed were shifted toward the red. In other words, the entire group of spectral lines recognizable as belonging to a specific element was shifted toward the red. This red shift needed explanation.

While many other astronomers preferred seeking the cause of the red shift in some property of interstellar space, Hubble insisted that it was caused by the Doppler effect, and this remains its universally accepted explanation. The Doppler effect is well known in acoustics: the pitch of a rapidly approaching ambulance siren is higher than it is at rest, and lower as it recedes. Light waves are subject to the same phenomenon. All the light emitted by an appoaching star is shifted slightly toward the blue end of the spectrum — the higher frequencies, like higher pitch in acoustics — and that of a receding star is shifted toward the red — lower frequencies, i.e., longer wavelengths.[50]

So if the correct explanation of the red shift is the Doppler effect, it follows that all the far-away stars and galaxies in the universe are receding from us. Not only that: Hubble found that the farther away they are, the faster they recede. The resulting curve, plotting the recession velocity of stars against their distance from us, is approximately a straight line: the speed with which a star moves away is proportional to its distance from us. The numerical value of the constant of proportionality is now known — Hubble's original estimate of its value was considerably higher — to be approximately 15 kilometers per second for every million light years. This is called Hubble's law — the ratio is known as the Hubble constant — which describes exactly how the universe as a whole is expanding.[51] According to the general theory of relativity it is space itself, with its curved, non-Euclidian geometry, that is expanding everywhere.

As to the question of the beginning of the cosmos and the form of all the energy in it, its whole history is now accepted by cosmologists to be governed by the general theory of relativity and quantum mechanics, including quantum field theory. The model universe so constructed is expanding, having started with a 'big bang'.

Hubble's law itself, of course, implies that, going backwards in time the universe contracts, and you can even calculate how long ago it was born as a point; its age is now estimated at about 13.75 billion years.[52] The 'big bang' is known as a singularity of the solution of the equations of general relativity, a point at which the equations break down. That's when time started and space rapidly began to expand, containing all the energy the universe would ever contain from its very beginning. To ask what happened before, or where the energy was before the big bang, is meaningless because there was no 'before.' Time and space did not exist until the universe was born.

From our discussion about the origin of all the energy the Earth receives from the Sun you might think that the earliest form of all energy was nuclear, but that is not so. It all started in electromagnetic form, that is, in the form of radiation. The tiny but rapidly growing baby universe was extremely hot and full of photons. In a sense the electromagnetic form may be regarded as 'naked' or pure energy in contrast to energy in the form of matter. The material form of energy emerged later.

When two photons collide, they can produce particles, as we have seen in Chapter 4, provided the total energy of the two photons is at least equal to the sum of the rest-masses of the created particles and all the other conservation laws are satisfied. To create an electron-positron pair, whose charges are equal and opposite so that their creation together satisfies the conservation of electric charge, requires a photon energy of $2\,mc^2$ or about 1 MeV, since the mass m of the electron equals that of the positron and each is about 0.5 MeV in energy units. The creation of neutrons or protons, or of heavier particles called baryons, requires about 2 GeV or more, since the law of baryon conservation requires that they be created in baryon-antibaryon pairs too. So the first particles created — in the material sense, with rest energy — were electrons and positrons, as well as protons and neutrons. Now, when a positron collides with a neutron, the electroweak interaction will sometimes turn the neutron into a proton, producing a neutrino and swallowing up the positron. So the positrons will eventually disappear. But what happened to all the antibaryons, including antinucleons? That is one of the great unsolved questions in physics. Are there regions of the universe populated by stars and galaxies made of antimatter? If a star were to encounter an antistar, they would both disappear in an enormous explosion, with nothing but radiation left over. Such events have never been seen by astronomers. So we are left with a puzzle, still unsolved.

After the creation of protons — hydrogen nuclei — and neutrons, the universe begins the process called nucleosynthesis: the breeding of the nuclei of all the chemical elements from the 'pure' energy present at the time of its creation.[53] We shall return to it soon.

In 1964 two radio-astronomers, Arno A. Penzias and Robert W. Wilson at Bell Telephone Laboratories in New Jersey made a strange discovery.[54] Their antenna had been especially constructed to keep the usual noise, the part of the 'static' accompanying all radio-transmissions originating from the electrical circuits in the antenna's amplifiers, as low as possible. And yet they kept on receiving a signal centered at a wavelength of 7.35 centimeters that did not seem to vary with the direction of their antenna; in other words, it seemed to come from everywhere. For a while, they suspected that the dirt left by a pair of roosting pigeons was responsible, but the birds were finally ruled out as the culprits when the antenna was thoroughly cleaned. So where did this radiation come from?

Radio-astronomers usually express noise they receive in terms of an 'equivalent temperature.' Here is what this means: as physicists well know, any given closed cavity will necessarily contain radiation whose strength at various wavelengths depends only on the temperature of the cavity. Penzias and Wilson described the radiation they observed as having an equivalent temperature of about 3.5 K (see Chapter 3 for the definition of the Kelvin temperature scale).

When astronomers and cosmologists read about this finding, they recognized that it was just about what George Gamow and his collaborators had predicted should be left over from the intense radiation in the early universe. If, after the completion of the process of nucleosynthesis, there had been any very hard (high-energy) radiation left, it would have been able to destroy all the newly created nuclei of the heavier elements. They had argued that, based on the known amount of hydrogen remaining in the universe, there was a specific peak in the frequency of the radiation left over. Furthermore, the enormous expansion of the universe since then had stretched space and hence red-shifted this peak so that it would now have an equivalent cavity temperature (the cavity in this case being the whole universe) of about 5 K.

The strange noise that Penzias and Wilson heard in their antenna was not caused by pigeon droppings but was the verification of an extremely fundamental process. It was nothing but the leftover wailing, so to speak, of the baby universe when it was done with transforming as much as possible of the pure energy it contained into matter in the form of nuclei of all the known elements.

Georgiy Antonovich Gamow, the man who had predicted the radiation leftover from the early universe, was born in 1904 in

Odessa (then in Russia but now in Ukraine). He was educated at the Universities of Odessa and Leningrad, where he became actively interested in the revolutionary new developments in quantum mechanics, pursuing research for his doctorate at the University of Göttingen as well as at the Cavendish Laboratory in Cambridge under Rutherford. The political atmosphere in the Soviet Union becoming increasingly oppressive, he and his wife, who was also a physicist, made two daring attempts at defection — one by kayak for about 160 miles across the Black Sea to Turkey, the other to Norway via Murmansk, both unsuccessful because of bad weather. Finally, after attending a physics conference in Brussels, they succeeded in emigrating to the United States in 1934. Gamow found a position at George Washington University, then at the University of California at Berkeley, and finally at the University of Colorado at Boulder. He died in 1968 in Boulder.

A highly imaginative physicist with fruitful original ideas and a playful sense of humor, Gamow also wrote many popular science books as well as the first book about theoretical nuclear physics. The paper that lent strong support to the big bang theory, initiating the notion of nucleosynthesis and allowing Penzias and Wilson to recognize their observed noise as the remnant radiation of the early universe, was written by him with the collaboration of Ralph Alpher. It was published, however, under the authorship of Alpher, Bethe, and Gamow (without even consulting Hans Bethe) simply because it tickled Gamow's funny bone to make a pun on the first three letters of the Greek alphabet: *alpha, beta, gamma*. It has been referred to as the α, β, γ paper ever since.

So we return to the workings of nucleosynthesis. The British astronomer Fred Hoyle proposed in the early 1950s that this process was responsible for generating all the chemical elements in the universe. The way this would work is by means of two steps employing fundamental processes: nuclear fusion and beta-decay of neutrons. The beginning is, of course, the creation of neutron-antineutron pairs directly from collisions of high-energy photons. Hydrogen nuclei (protons) are then formed by the beta-decay of neutrons (as well as directly by radiation as proton-antiproton pairs). The collision of a neutron with a proton will sometimes form a deuteron — the nucleus of ^{2}H — by fusion, and its subsequent fusion with another neutron will form tritium (see Chapter 3). The beta-decay of one of the neutrons of tritium then forms the nucleus of ^{3}He. After this the nucleus of ^{4}He — an alpha particle — is produced by fusion with another neutron.

This process continues, always adding a neutron to a nucleus already formed by fusion, thereby producing an isotope with an atomic weight increased by one, and the subsequent beta-decay of one of the neutrons in it will make it climb up the ladder of the periodic table of the elements by one unit. It is a very systematic step-by-step procedure capable of forming the nuclei of all the stable elements that are not produced later inside stars or by the occasional cataclysmic events of supernovae. The reason why many of the resulting nuclei fail to decay is, they are so tightly bound together that they are stable because their rest energy is less than the total energy of their decay products.

A few years after Hoyle's proposal, he and the American physicist William Alfred Fowler undertook the prodigious job of actually calculating the probability — using the tools of quantum mechanics — of each of the stepwise formations of all the elements in the periodic table by nucleosynthesis. The nuclei of the lighter elements were step by step generated in the relatively early universe, in the primordial nucleosynthesis, while those of the heavier ones — those from carbon on up — were predominently created in what was called stellar nucleosythesis, inside the stars that had by then been formed as the universe evolved, or by supernovae. This meant that Fowler and Hoyle calculated how many atoms of each element, from hydrogen to the heaviest ones, there should be in existence relative to one another, a number which astronomers call that element's *abundance* and which they have tabulated on the basis of observations of the strength of stellar spectra. By comparing the calculated results of Fowler and Hoyle with these observations, the theory of nucleosynthesis was confirmed and an important aspect of our idea of the development of the universe was verified. We therefore now know the details of how much of the energy in the cosmos was gradually tranformed from its initially pure form of radiation into matter making up the stars, the planets, and all of life.

The work of Fowler and Hoyle was rewarded in 1983 when William Fowler shared the Nobel prize in physics with Subrahmanyan Chandrasekhar. To everyone's astonishment Fred Hoyle was left out in the cold. What happened? Hoyle was a cantankerous man who had often gone out of his way to offend his colleagues, and he sometimes had very strange ideas. For years he had fought against the theory of the big bang, pushing instead his own notion of a 'steady-state universe' with no beginning and no end, in which there would be a continual creation of nucleons filling up the expanding cosmos. He had coined the term 'big bang' for the beginning of the universe to make fun of the theory he abhorred, but it remained in everyone's vocabulary without its intended irony. Paying little attention to what the laws of physics demanded, he stuck to his theory at the expense of his reputation, ultimately paying the price of losing the recognition of a genuinely fundamental contribution he had made.

There is, however, another form of energy in the cosmos. Recent astronomical observations seem to suggest an enormous store of invisible energy — called dark energy — in the universe, the prop-erties of which are at present quite unknown. The evidence for this mysterious energy, also called vacuum energy, comes from the remarkable, surprising observation by astronomers that the universe is not only expanding, the rate of this expansion is in fact accelerating! At the present time the nature of this dark energy — not only invisible but also subject to no other kinds of forces except gravity — is not at all understood. One possibility is that instead of an unknown kind of energy the accelerating expansion is due to that old bugaboo, Einstein's cosmological constant, which, as we have seen, he was only too happy to get rid of. Now this curious constant may need resurrection in some form. The final word on this, however, has not yet been written.

Notes

[49] That's how your brain estimates your distance from an object — based on the different angles at which it is seen by your two eyes.

[50] These shifts are very small unless the speed of the star is comparable to that of light.

[51] You should not conclude from this that there is something special about our position in the universe. The same law would hold for an observer situated anywhere else. All stars are receding from one another at this rate. Astronomers universally accept the so-called *cosmological principle*: there are no preferred places in the universe; the cosmos is essentially uniform in its general structure.

[52] For a nice and readable more detailed account of the age of the universe, see the book by David A. Weintraub, *How Old is the Universe?*

[53] For much more detail about what happened in the very early universe, see the book by Steven Weinberg, *The First Three Minutes.*

[54] Radio-astronomers observe the sky by means of large antennas receiving electromagnetic radiation of much longer wavelength than light, so-called radiowaves, including microwaves.

Epilogue

After going through the basic physics and chemistry of energy, its conservation, its various forms such as mechanical, electric, and chemical, its transformations from one into the other, and its storage and transport, we have learned that almost all the energy this planet uses for the maintenance of its modern civilization ultimately comes from the Sun. One of the few exceptions is nuclear power, which is also the source of the energy fueling the Sun. Another is geothermal energy, whose source ultimately is nuclear as well (see Chapter 3). Our reliance on the Sun extends not only to the fossil fuels but to the renewable ones like wind and water as well as whatever we are able to do with biomass, of course. After all, what makes the wind blow and produces the flow of rivers are temperature differences and mountainous elevation differences, both of which are driven by the Sun's heat, gravity, or both. Ultimately, therefore, almost all the energy used on the Earth originates from nuclear transformations, either fusion or fission.

When it comes to the question of the original form of all energy in the universe, however, we saw that it began in the 'pure' form of electromagnetic radiation. All the material form of energy, which makes up the galaxies, the stars, and the planets, was generated by pair creation, nuclear fusion, and beta-decay. The entire distribution of all the elements in the cosmos has been accounted for by these processes.

Are we on the Earth in any danger of eventually running out of energy, or running short of it? There is no reason at all to expect any such catastrophe. The time will certainly come when we run out of fossil fuels, long after access to it has become technically (or politically) so difficult that its price will have risen to prohibitive levels. At that point we will have to rely entirely on geothermal energy, renewable sources that depend on the constant flux of radiation from the Sun, and on nuclear power. Eventually we will of course also run out of uranium (or we fail to find an acceptable way of disposing of the radioactive waste), so that we can no longer rely on fission reactors. By that time we will surely be able to construct reliable fusion reactors, and their fuel, hydrogen, is inexhaustible. So we will never run out of the energy needed to sustain this civilization (at least certainly not before it is going to collapse for other reasons). From the scientific point of view, there is no reason to be pessimistic about the future availability of energy. Needless to say, there are other aspects of energy about which this book is silent.

Appendix: Research on Energy

In view of the importance of the ready availability of energy in the form most urgently needed, many individuals, organizations, and institutions, of course, are actively engaged in research and development on this subject, and the governments of many countries encourage and support it at their own laboratories as well as those of universities and private corporations. Such research has many features, not all of them scientific. Problems such as why and how to minimize the use of oil or natural gas and to find suitable substitutes, for example, have, in addition to environmental aspects, powerful political components. The Energy Department of the American government is actively promoting research of this kind in a variety of ways. Among these efforts is the establishment of a special agency called the Advanced Research Projects Agency — Energy, or ARPA-E, which funds and encourages both research and development of areas in the field of energy that promise high payoff. Here is a summary of such projects and of what they hope to achieve.

Battery Technology

There are many proposals to develop new and improved low-cost batteries by means new manufacturing processes, such as the following:

Lithium-ion batteries — using brown algae to make silicon eletrodes; metal-air ionic liquid (MAIL) batteries with the potential to increase the range of electric vehicles to distances approaching 1000 miles and to dramatically decrease their cost;

A novel silicon-coated carbon nanofiber paper for use in lithium-ion batteries, which can store four times more energy than existing technologies;

High-performance cathodes for lithium-air batteries (such batteries have extremely high theoretical energy densities approaching those of gasoline, due to the use of a high capacity lithium anode and oxygen from the air);

Grid-scale electrical storage using a flow-assisted rechargeable zinc-manganese oxide battery;

Batteries for intermittent renewable energy support on the electric power grid (such a proposed advanced multi-function energy storage (AMES) prototype would be expected to be highly scalable and provide multiple modes of operation);

A novel flow battery technology based on lead-acid chemistry that significantly reduces costs and extends battery life, using novel electrode materials that greatly increase the surface area available for chemical reactions, minimizing the amount of excess lead in the battery;

Hydrogen-bromine flow batteries for grid-scale energy storage, with the goal of developing a hydrogen-bromine (H_2-Br_2) flow-battery system for grid applications;

Rechargeable magnesium-ion batteries with high energy density; solid-state all-inorganic rechargeable lithium batteries;

A novel large format high-energy zinc-air flow battery for long range plug-in hybrid and all-electric vehicles;

An ultra-high energy lithium-sulfur battery able to power electric vehicles more than 300 miles between charges; an "all-electron battery", a completely new class of electrical energy storage devices for electric vehicles;

A robust and inexpensive iron-air rechargeable battery for grid-scale energy storage.

Other Energy Projects

Here is a sample list of other projects proposed to deal with the use, storage, or generation of energy:

Innovative building-integrated ventilation systems that promise much higher efficiency than conventional air conditioning systems;

Cyanobacteria designed for solar powered highly efficient production of biofuels;

A next generation flywheel energy-storage module that stores much more energy than current state-of-the-art flywheels, using a "flying ring": a lightweight hoop of co-mingled fiber composite with bonded magnetic materials mounted on the structure;

High yielding, low-input energy crops, using advanced plant breeding and biotechnology to develop new varieties of energy grasses (specifically, switchgrass, miscanthus, and sorghum) for use as feedstock for the production of biofuels;

A novel wind turbine that delivers significantly more energy per blade diameter than existing models, enabling the deployment of wind turbines in a wider range of locations, including urban environments;

A thermal-mechanical drilling technology that will dramatically increase drilling rates in penetrating ultra-hard crystalline basement rocks relative to conventional drilling technologies, as the use of geothermal energy has been hindered because conventional drill bits penetrate these rocks slowly and wear down quickly;

Innovative compressed air energy storage technology that could accelerate the integration of renewable electricity resources, particularly wind, into the grid;

A system for recovering waste heat in automobiles, utilizing shape memory alloys (SMAs), which are deformed by heat and return to their original form at cooler temperatures;

A bacterial reverse fuel cell by engineering a bacterium to absorb electrical current as an input and convert this energy into chemical energy by fixing carbon dioxide in the form of a biofuel, specifically octanol;

A novel film coating for windows consisting of a solid-state electrochromic material on plastic substrates that will significantly reduce heat and energy loss in all types of buildings;

An airborne wind turbine, using tethered, high-performance wings outfitted with turbines, power being extracted from this motion by the wing-mounted turbines and transmitted to the ground through an electrically-conductive tether (because the wing is not constrained to rotate about a hub, it can sweep a much larger section of the sky than a conventional wind turbine and fly at a higher altitude where the wind is both stronger and more consistent);

A phononic heat pump, i.e., a refrigerator that pumps heat using sound waves; ammonothermal bulk gallium-nitride crystal growth for energy-efficient solid-state lighting devices;

A novel biological conversion of hydrogen and carbon dioxide directly into biodiesel;

Adaptive turbine blades: blown wing technology, wings that can be dynamically adjusted by circulation control to maximize power under a wide range of wind conditions for low-cost wind power;

An advanced electrochemical energy storage device that incorporates a regenerative electrolyzer and fuel cell and an alkaline membrane;

To exploit a novel water oxidation catalyst discovered at MIT that employs earth-abundant elements to generate hydrogen and oxygen from tap water or clean sea water;

To engineer *shewanella* bacteria to produce higher levels of hydrocarbons from carbon dioxide, thus removing carbon dioxide from the atmosphere and producing hydrocarbon biofuels.

Illustration Credits

Figure 1.1:   From Murphy, *Inventions*, page 29.

Figure 1.2:   The first image: created by the United Kingdom Government, Science Museum, 1958 (from Wikimedia Commons); the second image: also Wikimedia Commons; the third image: a photograph of TGV Duplex at Gare de Lyon (Paris), taken on 31 August 2005 by Sebastian Terfloth.

Figure 1.3:   Photograph by Ben Paarmann, London, UK. (via Wikimedia Commons).

Figure 1.4:   From www.virtualtourist.com/travel/Asia/India/Mumbai.

Figure 1.5:   From www.tpub.com/content/construction/14264/css/14264 42.htm.

Figure 1.6:   From *Scientific American*, July 1984, page 125.

Figure 1.7:   From Ohanian, *Principles of Physics*, page 406.

Figure 2.1:   From Library of Congress, Prints & Photographs Division, Prokudin-Gorskii Collection.

Figure 2.2:   Self generated by means of Adobe Illustrator.

Figure 2.3:   From tpub.com/content/construction/14264/css/14264 42.htm.

Figure 3.1:   After the image in http://csep10.phys.utk.edu/astr162/lect/energy/cno-pp.html.

Figure 4.1:   This image appears in most elementary physics books.

Figure 4.2:   Google Images, *aurora borealis.*

Figure 4.3:   This illustration was produced by Luc Viatour (www.lucnix.be).

Figure 5.1:   From the web, but a very common photograph.

Figure 5.2:   From Hecht, *Physics: Calculus,* page 727, Fig. 19.5.

Figure 5.3:   From *Science,* 24 June 2011, page 1494.

Figure 5.4:   From *Science,* 24 June 2011, page 1494.

References and Further Reading

Arcoumanis, C., and T. Kamimoto, *Flow and Combustion in Reciprocating Engines.* Berlin, Heidelberg: Springer Verlag, 2009.

Bernstein, Jeremy, *Hans Bethe, Prophet of Energy.* New York: Basic Books, 1980.

Brown, T. M., "Resource letter EEC-1 on the evolution of energy concepts from Galileo to Helmholtz," *American Journal of Physics* 33 (1965), 759–765.

Fletcher, Seth, *Bottled Lightning: Superbatteries, Electric Cars, and the New Lithium Economy.* New York: Hill and Wang, 2011.

Hecht, Eugene, *Physics: Calculus.* Pacific Grove, CA: Brooks/Cole Publishing Co., 1996.

Hecht, Eugene, "How Einstein confirmed $E_0 = mc^2$", *American Journal of Physics,* 79, June 2011, page 591.

Hiebert, E. N., *Historical Roots of the Principle of Conservation of Energy.* Madison, Wisc.: Ayer Co.Publ., 1981.

Isaacson, Walter, *Einstein: His Life and Universe.* New York: Simon & Schuster, 2008.

Miller, David Philip, *James Watt, Chemist: Understanding the Origin of the Steam Age.* London: Pickering & Chatto, 2009.

Murphy, Glenn, *Inventions.* New York: Simon and Schuster Books for Young Readers, 2008.

Newton, Roger G., *Galileo's Pendulum: From the Rhythm of Time to the Making of Matter.* Cambridge, Mass.: Harvard University Press, 2004.

Newton, Roger G. *From Clockwork to Crapshoot: A History of Physics.* Cambridge, Mass.: Belknap Press, 2007.

Ohanian, Hans C., *Principles of Physics.* New York: W.W. Norton & Co., 1994.

Palz, Wolfgang, *Power for the World: The Emergence of Electricity from the Sun.* Singapore: Pan Stanford Publishing Pte. Ltd., 2011.

Planck, Max, *Treatise on Thermodynamics.* English ed., New York: Dover, 1945.

Rosen, William, *The Most Powerful Idea in the World: A Story of Steam, Energy, and Invention.* New York: Random House, 2010.

Smil, Vaclav, *Energy in World History.* Boulder: Westview Press, 1994.

Smil, Vaclav, *Energies: An Illustrated Guide to the Biosphere and Civilization.* Cambridge, Mass.: The MIT Press, 1999.

Smil, Vaclav, *Energy: A Beginner's Guide.* Oxford: Oneworld Publications, 2006.

Smil, Vaclav, *Energy in Nature and Society: General Energetics of Complex Systems.* Cambridge Mass.: MIT Press, 2008.

Smith, Crosbie, *The Science of Energy: A Cultural History of Energy Physics in Victorian Britain.* The University of Chicago Press, 1998.

Weinberg, Steven, *The First Three Minutes: A Modern View of the Origin of the Universe.* New York: Basic Books, 1993.

Weintraub, David A., *How Old is the Universe?* Princeton University Press, 2011.

Index

abundance, 82
age of the Earth, 36
alpha
    particle, 41, 49
    radioactivity, 34, 49
    rays, 34
Alpher, Ralph, 81
alternating current, 72
Ampère, André Marie, 23, 72
angular momentum, 9
angular velocity, 7
anode, 66
antibaryons, 79
antimatter, 79
Aristotle, 6
ARPA-E, 87
asphalt, 63
atomic bomb, 19
aurora borealis, 55
automobile, 64

Barber, John, 28
batteries, 87
Becquerel, Henri, 34
beta decay, 54
beta radioactivity, 35
beta rays, 34
Bethe, Hans Albrecht, 41, 43, 81
big bang, 78, 81, 83
biomass, 62
Bohr, Aage, 48
Bohr, Niels, 35, 47, 48, 73
Born, Max, 47, 49

breeder reactor, 38
Brownian motion, 16

caloric, 11
carbohydrates, 31, 61
carbon dioxide, 28
Carnot, Nicolas Léonard Sadi, 14
cathode, 66
cavity radiation, 80
chain reaction, 37
Chandrasekhar limit, 44
Chandrasekhar, Subrahmanyan, 83
chemical bonds, 51
chemical combustion, 40
chemical energy, 27
chlorophyll, 30, 61
CNO cycle, 41
CO_2, 28
coal, 61
coherent, 71
combustion, 27
comet, 7
conservation laws, 9
conservation of energy, 13, 34
corona, 57
Cosmic Egg, 76
cosmological constant, 76, 84
Cosmological Principle, 84
creation of particles, 19
critical mass, 38

Daimler, Gottlieb, 28
dark energy, 83

degrees kelvin, 40
deuterium, 39
deuteron, 39, 82
Diesel, Rudolf, 28
Dirac, Paul, 47, 49
Doppler effect, 77
dynamo, 23

$E = mc^2$, 52
Eddington, Arthur Stanley, 17
Edison, Thomas, 21, 67, 74
Einstein, Albert, 16, 18, 26, 47, 49, 71
electric generator, 23
electric motor, 23
electrical battery, 65
electrical engine, 22
electricity, 21
electrodes, 65
electroluminescence, 22
electrolyte, 65
electromagnetic
    field, 24
    force, 24
    induction, 23
    radiation, 24
electron volt, 58
electroweak theory, 54
elementary particles, 52
engine, 31
Euler, Leonhard, 9
eV, 58
evolution, Darwin's theory of, 36
excited atom, 55
exclusion principle, 58
exoergic, 31
exothermic, 31

Faraday, Michael, 23, 24, 72
Fermi, Enrico, 35, 36, 38, 54
Feynman, Richard, 52
first law of thermodynamics, 13
fissions, 37
fluorescence, 22
fluorescent bulbs, 22
flying ring, 89

flywheel, 64
fossil fuel, 61, 64, 86
Fowler, William Alfred, 82, 83
friction, 10, 11
Frisch, Otto, 37
fuel cell, 69
fusion, 39, 81
fusion reactors, 40, 86

Galileo, 6
Galvani, Luigi, 67
gamma
    radiation, 41
    rays, 34
Gamow, George, 50, 80
gas, 63
gas turbine, 28
gasoline, 64
geothermal energy, 35
GeV, 58
Glashow, Sheldon, 54
ground state, 48

H_2O, 27
Hahn, Otto, 36
half-life, 34, 50
halogen bulbs, 21
hard photons, 57
heat
    kinetic theory of, 11
    mechanical equivalent of, 13
heat death of the universe, 15
Heisenberg, Werner, 38, 47, 49
helium, 58
Helmholtz, Hermann von, 12
Henry, Joseph, 23
Hero engine, 1
Hertz, Heinrich, 24
Hilbert, David, 10
Hoyle, Fred, 81
Hubble constant, 78
Hubble's law, 78
Hubble, Edwin, 77
Huygens, Christiaan, 1
hydrocarbons, 27, 30, 63

hydroelectric power, 23
hydrogen, 27
hydrogen bomb, 39
hydrogen, liquid, 68

incandescent light, 21
industrial revolution, 1, 62
infra-red, 57
internal combustion engine, 28, 64
invariance, 10
isotope defined, 45

Joule, James Prescott, 12

Kelvin, 36
kerosene, 64
KERS, 65
kilowatt hours, 22
kinetic energy, 6
        rotational, 7

laser, 70
Lavoisier, Antoine Laurent, 27
LED, 22
Lemaître, George Edouard, 75
Lenard, Philipp, 25
Lenoir, Etienne, 28
light-emitting diodes, 22
line spectrum, 55
lithium-ion batteries, 88
locomotive, 2
Łukasiewicz, Ignacy, 64

Manhattan Project, 19
maser, 73
mass, energy equivalent of, 19
Maxwell's laws, 24, 47
Maxwell, James Clerk, 24
Maybach, Wilhelm, 28
Mayer, Robert, 11
Meitner, Lise, 37
Mendeleyev, 51
MeV, 58
moment of inertia, 7
momentum, 9

neptunium, 37
neutrino, 35
neutrino oscillations, 42
neutron, 39, 54
neutron beta-decay, 81
neutron star, 44
Newton, Isaac, 7, 16
Noether's theorem, 10
Noether, Amalie Emmy, 9
nuclear
        energy, 36, 38
        power, 85
        reactions, 19
        reactor, 38
nucleon, 49, 54
nucleosynthesis, 79–81
        primordial, 82
        stellar, 82
nucleus, 47
nucleus of the atom, 34

oil, 63
oil refinery, 64
oil well, 64
oxygen, 27

paddle wheels, 7
pair creation, 79
Papin, Denis, 1
parallax, 76
particle collisions, 52
Pauli, Wolfgang, 35
pendulum, 7, 51
Penzias, Arno A., 79
periodic table, 51
perpetuum mobile, 13
        of the second kind, 15
petroleum, 63
petroleum distillation , 64
phlogiston, 27
photoelectric effect, 25
photon, 26, 51
photosphere, 57
photosynthesis, 30, 61, 62
photovoltaic effect, 26

pile, 38
Planck, Max, 17, 26, 48
plutonium, 37
positron, 41
potential energy, 6
power, 22
PPII chain, 43
PPIII chain, 44
pp cycle, 41
Priestley, Joseph, 27
Primeval Atom, 76
proton, 39
pumped storage, 65

QED, 49, 51
quanta, 26
quantized, 26
quantum electrodynamics, 49
quantum field theory, 52
quantum jumps, 55

red shift, 77
relativity, general theory of, 17
relativity, theory of, 16, 18
renormalization, 52
respiration, 27
rest energy, 19
rotation, 7
Rutherford, Ernest, 33, 47, 81

Salam, Abdus, 54
Schrödinger, Erwin, 47, 49
Schwinger, Julian, 52
second law of thermodynamics, 14
shape memory alloys, 90
Shapley, Harlow, 75, 76
shewanella bacteria, 91
singularity, 78
soft photons, 57
solar neutrino puzzle, 42
solar panel, 26
solar power plants, 26
spectrograph, 55
steady-state universe, 83
steam engine, 1, 2
stimulated emission, 71
Strassmann, Fritz, 36

Sun, 85
sunlight, 24
sunshine, 57
superconducter, 74
superconductivity, 21
supernova, 44
symmetry, 10

Tesla, Nicola, 74
Theophrastus, 62
thermodynamics, 5
thermonuclear reaction, 39
Tomonaga, Sin-itiro, 52
torque, 7
transformer, 72
transmission lines, 72
transuranic elements, 37
Trevithick, Richard, 3
tritium, 39, 82
triton, 39
tunneling, 50

ultraviolet, 57
universe, size of, 76
uranium, 36, 86

vacuum energy, 83
variable stars, 76
vis viva, 6
Volta, Alessandro, 67
voltage, 22

waste heat, 15
water, 27
Watt, James, 1, 2
watts, 22
weak interaction, 54
Weinberg, Steven, 54
Weizsäcker, Carl Friedrich von, 41
whale oil, 64
white dwarf, 44
Wilson, Robert W., 79
wind turbine, 89, 90
windwheel, 1
work, 6, 7

zero-point energy, 51